SUSTAINABLE WORLD SOURCEBOOK

CRITICAL ISSUES • INSPIRING SOLUTIONS • RESOURCES FOR ACTION

Compiled by the Sustainable World Coalition

For the children of all species

TESTIMONIALS

DEDICATION

Indigenous human cultures around the world celebrate the sacredness of nature. They have developed richly diverse spiritual ways of life that honor the laws of nature, which rule this beautiful world.

We, the divine humans upon sacred Earth, now find ourselves at a prophetic crossroads with our profound planetary destiny. Each of us brings to this immense moment all of our beloved ancestors, and seven generations of those not yet born. We focus our lives and resources to foster and perpetuate a nature-friendly planetary community in the arenas of religion, commerce and politics.

All women—the life-givers—are Mother Earth's feminine presence. Feminine leadership and values are essential now to balance the masculine.

Thus we renew and celebrate our sacred relationships with mother nature, our most precious gift. We do so for all children of all creatures—to ensure that they will inherit a healthy, happy and peaceful world from us.

Yata Hey—with love from my heart.

—**Chief Sonne Reyna** (Lipan Apache-Yaqui)

Bluebird Woman

FOREWORD BY PAUL HAWKEN

There is a possibility that the world will see again. At this moment, global civilization is blind to the peril and possibility that resides in forces at play in society and the environment. To rectify that, there is an essential and indispensable course of action, a systematic call from each village and neighborhood and family to others, to become literate in the social and environmental challenges we face.

Aristotle said that genius is metaphor, but there is as yet no metaphor (or genius) that can adequately describe or encompass this moment in time because there is no precedent. The terms crossroads, turning point, watershed, and moment of truth do not address this juncture for these are civilizational times and they require each of us to reimagine what it means to be a human being. Critical to this act of newly conceiving who we are and what we should be doing is literacy. The *Sustainable World Sourcebook* is a door to a newly born and critically important literacy of the world around us.

Literacy is the first step to dialogue, conversation, and collaboration. If you cannot read, books look like cooking fuel. You may see a bird as gossamer and feathers, but it is also a creator of forests and meadows, flying with its small sac of undigested seeds. We can see the world as doomed and fatally flawed, or we can see every trend and statistic as the possibility of transformation. We can see ourselves as fortunate and separate from the suffering of others, or we can see that our bounty rests heavily on the shoulders of unknown people. Concomitant to our good fortune is the responsibility to create a world of equals, not just a nation of equals.

As stunning and paralyzing as is the data about climate change, the possibilities that are emerging from the imagination and concerns of humankind are equally stunning. In energy alone, we are tasked with reducing carbon-based energy by 80% in the next 25 years if we are to prevent temperature from rising more than 2 degrees Centigrade.

Yes, the world uses 84 million barrels of oil per day. But, a square meter of the Earth receives in one year the same amount of energy as a barrel of oil. The US, a profligate user of energy, has about 4,000 times more solar energy than its annual electricity use. For the world, that figure is 10,000 times, which means that if 1% of the world's land were used for solar photovoltaics, we would have 10 times the amount of energy needed. And a new branch of thin-film solar technology is nearly at grid parity, meaning that a solar panel can produce electricity at costs commensurate with coal and gas. This example is repeated in hundreds of areas of social and technical innovation. Humanity knows what to do, once it knows what needs to be done.

To make a change this broadly this quickly is unheralded. It calls for a mobilization where human beings are not rallying to defeat an enemy, but organizing to support each other. It is not just our energy and resource profligacy that is devastatingly expensive; our divisions, dislikes, and antagonisms are equally unaffordable. Coming together in communities of empowered people, to quote former President Clinton, is the work at hand, and this book and the extraordinary experience of the Awakening the Dreamer Symposium are critical tools in humankind's journey to a world that is conducive to and protective of all life.

Paul Hawken is an environmentalist, entrepreneur, journalist, and best-selling author. He has dedicated his life to sustainability and changing the relationship between business and the environment. Paul resides in Northern California.

TABLE OF CONTENTS

Publisher's Preface

What an amazing time to be alive! It's truly an historic moment in human history—one in which we will very soon be forced to deal with the mess we've made of things on the planet and either learn how to work together to clean it up or else go the way of other human cultures that went extinct when they didn't heed their environmental desecration (e.g., Easter Islanders).

There is a huge awakening moving through humanity, a recognition of what is at stake, and ways that we may all contribute to what Joanna Macy calls The Great Turning. Indeed, we all have the chance to live very meaningful lives of purpose and contribution, which is a great blessing and opportunity. And this starts with learning about where things stand with the big issues we are now facing as humanity, and what we can all do about it.

This *Sourcebook* is designed to inspire you to gain an essential understanding of the challenges facing us all, and to make the connection between our daily life choices and our impact on the world. The *Sourcebook* brings clarity to the major issues of our time, presents some of the best solutions, and provides guidelines for effective personal and collective action. Our purpose is to contribute to the creation of a thriving, just and sustainable world for all—all people and all other species.

In recent years, increasing numbers of people and organizations are addressing environmental devastation, resource depletion, species extinction, and extreme wealth inequity by exposing the truth, advocating for change, and implementing solutions—there is a great awakening under way.

According to current data, we are rapidly nearing a point of no return for saving much of what we know and love on our beautiful planet. There is no time to waste! Restoring balance and sustainability across the globe will take our best minds and a critical mass of citizens who are not just concerned but committed to making the changes required.

We know from past crises that as things continue to heat up and get worse, we will begin to experience real humility, rethink what's important, and mobilize for action as never before. We will find ourselves working together for the common good in ways we only now dream of. That vision gives me great hope and inspiration—we *can* do this!

So where do we begin?

STEP 1: GET INFORMED. Find out about the issues both global and local, the proposed solutions, and how we as individuals can contribute to creating a thriving, just, and sustainable world.

STEP 2: GET ENGAGED. Open your heart. Engage your creativity, your passion, your dreams. Deepen your connections to the Earth and one another. Act on what you know and learn. Get involved with improving your community. Live a life with real purpose. Challenge your friends to do the same—we're all in this together.

Thank you for reading this book and doing your part as an engaged global citizen... Godspeed!

Vinit Allen
Director, Sustainable World Coalition

A World of Thanks

Claus Wawrzinek

We appreciate the individuals, networks and multitude of organizations doing the important, leading edge work that inspired and informed the process of compiling the *Sourcebook*. Our gratitude goes to all those whose advice, assistance talents and commitment to creating a better world made this book possible.

CONCEPT & PUBLISHER: Vinit Allen PROJECT DIRECTOR: Steve Motenko MANAGING EDITOR: Sara Stroud EDITOR: Elizabeth Grossman ASSOCIATE EDITOR: Dazzia Szczepaniak RESEARCHERS: Cristina Parks, Kate Forrest, Katie Aiello, Harmony Simes, Karen Motenko-Neal, Rebecca Snyder, Jimmy Luong, Michael Leao, Olivia Feinstein, Beth Vanden Heuvel, Madeleine Fisher, Teresa Fukuda, Jennifer Tang, Michelle Claudio, Chenye Liu, Hannah Han, Christie Zhang, Robert Edgar, Gabe Falzone, Monica Bunnay, Lei Kang, Molly Ehlers, Suzy Karasik, Sue Staropoli WRITERS: Lynne Michelson, Jody Bol, June Holte, Jim Embry, Trathen Heckman, Ellen Augustine CONTENT INPUT: David Korten (YES! Magazine), Robin R. Milam (Global Alliance for the Rights of Nature), Linda Sheehan (Earthlaw Center), Martin Bourque (the Ecology Center), Gregory Mengel and the Pachamama Alliance Social Justice Team PHOTOS: Chris Jordan, Dazzia Szczepaniak, Gloria Garrett, Roberta Vogel, Emmanuel Dyan, Claus Wawrzinek, Jim Maragos, Russell Watkins, Neil Palmer, Richard Ling, Robin Pierro, Lydur Skulason, Michael Mees, Pierre Holtz, Shehzad Noorani, Jonathan McIntosh, Erin Whittaker, Joseph Woodard, Lillian Dignan, Tom Chance, Keith Bacongco, Corey Templeton, Ajay Tallam, Llee Wu, Natalie Maynor, Scott Sporleder, Eddie Thornton, Amanda Rhode, Leah Beck CARTOONS: Brian Narelle ADDITIONAL SUPPORT: Dazzia Szczepaniak, Renée Soule, Ann West, Rua Necaise, Kevin Ohmsman, Rajyo Markman FUNDING: Allen Family Foundation.

From the Managing Editor

Gloria Garrett, gloriousjourneyphotography.org

Welcome to the new edition of the *Sustainable World Sourcebook*. We trust you will find it to be a valuable tool to support you in your personal journey of contributing to a thriving, just, sustainable world.

The *Sourcebook's* new i-book version, complete with embedded videos, adds to its value as a living, breathing document. Throughout both the print and e-versions, you'll find a host of links to serve as jumping off points for your own exploration.

Although we provide a global context, the primary distribution of the printed version of the book will be in the United States—the country with the largest ecological footprint and largest economy, and thus the country with arguably the greatest responsibility to address our collective challenges.

The *Sourcebook* is a forum rather than a definitive treatment or position. We aimed to find and document the most credible sources available. We've included an Endnotes section, so you can check out our source material. Also included is a robust Resource Directory with many useful organizations, programs and information.

Our researchers and contributors are not responsible for the content of this collaborative project, the compilation of information, or views expressed. If you spot errors or omissions, please help us fix them! We sincerely hope you will visit our website, swcoalition.org, to interact with us and check out our programs and courses.

Most of all, this *Sourcebook* is a labor of love. Many people volunteered untold hours, exploring constantly evolving content as the sustainability movement expands around us. The writers, researchers, editors, and reviewers include both established and emerging leaders and experts. Volunteers include practitioners, graduate students from sustainability/green institutions, and Awakening the Dreamer Symposium facilitators. Our gratitude extends to Vinit Allen, for his vision, funding, distribution, and editorial partnership in this project.

Sincerely, *Sara Stroud*

Part One: SETTING THE CONTEXT

Now you begin to see that your dream is a nightmare… All you have to do is change the dream… You need only plant a different seed, teach your children to dream new dreams.

—Elder of the Shuar tribe, Ecuador

In the summer of 1995, a small group of Westerners answered a call for partnership from one of the world's most remote indigenous cultures, the Achuar of the Ecuadorian rainforest.

Thus was born The Pachamama Alliance *(pachamama.org)*. *Pachamama* is a word from the Andean language Quechua; its meaning encompasses the sacred presence of the Earth, the sky, the universe, and all time, although it is often translated as "Mother Earth."

This partnership supports a federation of tribes working together to protect over two million acres of pristine rainforest. Its achievements include something unprecedented: Nature has been granted rights in Ecuador's constitution.

Yet all along, the Achuar suggested that a critical aspect of the work must be done not in the rainforests, but among industrialized societies in the global North. That work: changing the "dream" of the modern world.

So in partnership with dozens of organizations and individuals, The Pachamama Alliance created the "Awakening the Dreamer, Changing the Dream"

Symposium.[1] Since 2005, more than 3,500 trained volunteer facilitators have been taking the Symposium around the world, sharing it with groups large and small in more than 60 countries, on six continents, in many languages.

"From the very beginning, our indigenous partners told us that … if we really wanted to protect their lands permanently, we would need to go to work in our part of the world, and, as they put it … to change the dream of the North, a dream rooted in consumption and acquisition without any regard to the consequences for the natural world, or even for our own future."

—Lynne Twist, co-founder, The Pachamama Alliance

The *Sustainable World Sourcebook* and the Awakening the Dreamer Symposium share a common purpose: to bring forth an environmentally sustainable, spiritually fulfilling, and socially just human presence on this planet. Although these may appear to be three separate missions, they are actually three facets of one interconnected whole.

Millions across the globe who pursue this three-part mission are embarking on what eco-philosopher Joanna Macy describes as a new relationship with ourselves, each other, and all life on this planet. We are changing the "dream of the modern world."[2]

To set the context for everything that follows in these pages, we open with a summary of the Awakening the Dreamer Symposium, exploring both the root causes of our global crises and the promising possibilities for the future of our planet. We will discover that the critical global issues we face are symptoms of our consciousness, rooted in our cultural story. This is the place to start.

The issues outlined below are explored in greater detail in later chapters of the *Sourcebook*.

"The most remarkable feature of this historical moment on Earth is not that we are on the way to destroying the world—we've actually been on the way for quite a while. It is that we are beginning to wake up, as from a millennia-long sleep, to a whole new relationship to our world, to ourselves and to each other."

—Joanna Macy, eco-philosopher, author

WHERE ARE WE?
A status report on the health of our planet

Environmental Sustainability

As a result of climate change alone, experts say, the world as we know it may become unrecognizable within two generations—or less. But it's not just about climate. In every natural domain, the support systems for life on Earth are under very severe stress.[3]

- 60% of the ozone layer has been lost in 50 year
- 70% of the world's original forests are gone
- 30% of the world's cultivatable land has been lost in the last 40 years
- 90% of all large fish are gone from the oceans

"Sustainability is the ability of the current generation to meet its needs, without compromising the ability of future generations to meet theirs."

—United Nations, 1987[4]

As a global civilization, we are now using 50% more natural resources than the Earth can regenerate, thus eroding the natural capital that life depends on. If everyone lived as North Americans do, we would need five Earths to sustainably maintain our way of life. If everyone lived as Europeans do, three Earths. China and India are just under the one-Earth consumption level, but with rapid industrial development in those countries, that proportion is increasing steadily.[5]

This overconsumption of the Earth's bounty is accompanied by a catastrophic loss of our planet's biodiversity. The World Wildlife Fund's Living Planet Index, which tracks more than 9,000 populations of more than 2,500 animal species, has revealed a 28% decline in biodiversity health around the world since 1970.[6]

"We lament the passing of the people we love, and our pets … but how to lament the permanent loss of a mode of life?

Slum housing in Ahmedabad, Gujarat, India

That's beyond most of us, because we haven't deepened our hearts in a way that would make possible the grief that is wanting to be felt."

—**Brian Swimme**, *author, professor, cosmologist*

Social Justice

If you have food in a refrigerator, clothes in your closet, a bed to sleep in, and a roof over your head, you are better off, materially, than 75% of people on this planet.[7]

No one would assert that we have achieved a socially just human presence on the planet. More than 1 billion people worldwide live in urban slums, partly as a result of massive migrations to cities from environments that have sustained people for generations. Many of these slums lack safe structures and dependable sanitation.[8]

Almost half the world's population lives on less than $2 a day.[9] The gap between rich and poor has been widening rapidly everywhere since the 1980s. This growing gap is a disturbing trend that exacerbates numerous social problems.

And environmental crises are also exacerbating social problems. Climate destabilization, resource depletion, pollution, and habitat destruction are al-ready beginning to have devastating impacts on millions—especially in socioeconomically disadvantaged parts of the world.

The movement to address this disparity is called environmental justice. It reflects a deep, innate human longing for justice and equity, and links environment, social justice and economics.

"Environmental justice is the belief that no community should have to bear the brunt of a disproportionate amount of environmental burdens and not enjoy any environmental benefits. But right now race and class are the best indicators as to where you'll find the good stuff—like parks and trees—and where you'll find the bad stuff—like waste facilities or power plants."

—**Majora Carter**, *founder, Sustainable South Bronx*

Spiritual, Psychological and Emotional Impact

What about our personal sense of wholeness and fulfillment? The costs of our modern worldview are evident. Study after study shows that beyond a baseline of subsistence, material gains do not lead to greater happiness or spiritual fulfillment. Our possessions are scant comfort when we lack a sense of belonging.

"There is a great loneliness of spirit today. We're trying to … cope in the face of what seems to be overwhelming evidence that who we are doesn't matter; that there is no real hope for change; that the environment is deteriorating rapidly, and increasingly, and massively. Meanwhile, we're yearning for connection with each other, with ourselves, with the powers of nature, with the possibilities of being alive.

"When that tension arises, we feel pain, we feel anguish at the very root of ourselves, and then we cover that over, that grief, that horror, with all kinds of distractions—with consumerism, with addictions, with anything we can use to disconnect. But if we can open to that grief, that pain, there's a possibility of embracing [it] in a way that it becomes a strength, a power to respond."

— *John Robbins, author,* Diet for a New America

HOW DID WE GET HERE?
Cultural Drivers and Unexamined Assumptions

The bad news is that human decisions and human behavior created most of our current challenges, worldwide. We did this to ourselves! But the bad news is also the good news—if our actions created this outcome, then different actions can create different outcomes.

It could be said that in the modern world we are living in a kind of trance, what some indigenous people would call the "dream of the modern world." The late eco-theologian Thomas Berry referred to our age as a period of "technological entrancement"—reflecting an ill-defined "dream of progress." Yet it's so ingrained in our culture that we don't question it.

Berry scholar and poet Drew Dellinger describes it like this: "We think we're behaving rationally—creating jobs, increasing gross domestic product, thinking we're on a kind of logical economic course—but actually we're heading toward our destruction."

The reason, Dellinger explains: "Western civilization creates and perpetuates a radical separation between the human world and the natural world. We give all rights to humans and none to the natu-

ral world." In doing so, we forget that the "human world" *depends entirely* on the health of the natural world.

This trance is our current worldview—a point of view so pervasive that it is invisible to us. It is held in place by a set of beliefs and unexamined assumptions, like glasses we've worn so long we don't realize we're looking through them.

The notion that humans are separate from nature is one such unexamined assumption. In fact, it's the central one—the one that underlies all the others, such as:

- Nature is a resource for human use

- The "resources" of nature are unlimited

- A healthy society depends on a growing economy

- When we buy something, the price we pay reflects the full cost of making it.[10]

- There is a place called "away" where we can throw things. (As activist Julia "Butterfly" Hill asks, "Where is 'away'?")

- We have disposable resources, disposable species, and disposable people

- One individual can't make a difference

Trying on Other Worldviews

People take actions appropriate to the way they see the world. If our assumptions lead us to take actions that produce negative outcomes, then different assumptions should lead to different actions and thus different outcomes.

What alternative assumptions might create a healthier, more accurate depiction of planetary reality? Exploring other possible worldviews can help us gain perspective on what else is possible.

Many people on this planet, including indigenous cultures that have lived sustainably on Earth for thousands of years, are not caught in the "dream of progress." They have a very different way of seeing the world and their place in it.

"We use another terminology, called 'Mitakuye Oyasin,' which is, 'All My Relations.' We recognize that we are

Native Americans were (and often are) natural custodians of the land, using sustainable management practices

human being to human being. And I'm also related to the animals, to the plants, even the micro-organisms.... Somehow, industrialized society has not caught up with itself to really appreciate and respect what indigenous peoples have to offer, but it's something that's very important — that's going to save the planet."

—**Tom Goldtooth**, *executive director, Indigenous Environmental Network*

The contrast here is striking and informative: The worldview that has come to prominence over recent centuries is that the world is like a huge machine made up of separate parts, like a big clock—a mechanical, rather than organic, model of life. But this worldview is missing something profound, and wreaking widespread devastation.

No wonder we've been creating such havoc—our destructive behavior and its unintended consequences are the result of an inaccurate worldview. Our actions result from a cultural perspective we can now see as misguided and limited. We don't need to assume we are somehow innately flawed. We have simply been mistaken in how we see the world. We can learn to see it better.

And we are indeed learning. We in the modern world are coming back around to what many indigenous cultures have known all along: The world is interconnected and interrelated. Gradually, but steadily, we are waking up from the nightmare—the "dream of the modern world."

Transforming the Cultural Story— A New Story for All of Humanity

Our shared worldview, and the host of unexamined assumptions that comprise it, can be called our "cultural story." At this time in history, our story is undergoing a transformation—a transformation no less profound than the shift from believing the Earth was at the center of the universe to understanding that the Earth revolves around the sun.

Imagine the clash between those who embraced the new understanding—that the sun is the center of the solar system—and those who clung to the belief that the Earth was the stationary center of everything. This is analogous to the clash evident everywhere today. It's no wonder that huge numbers of people are, for example, denying the human sources of climate change despite overwhelming evidence. As in Copernicus' time, people cling to their worldviews with sometimes irrational ferocity. To do otherwise is, for some, simply too frightening.

But the current "new story" of interconnectedness, consistent both with modern science and with centuries-old indigenous understanding, is beginning to shape the consciousness on our planet, opening the possibility of a future that is not merely an extension of the past.

"A profound wisdom is at work in the Universe.... As we move into this understanding, we have a new identity of ourselves as cosmological beings. We're not just Americans, we're not just French, we're not any small category. We are the Universe in the form of a human.

"Everything is part of this. ... No matter what being we're talking about on the planet, we are related. We're related in terms of energy. We're related in terms of genetics. We're all a form of kin.... It is a massive change in human consciousness."

—**Brian Swimme**, *cosmologist*

This new understanding changes everything. In the emerging story of universal connection, behaviors that would previously have been considered an assault on the Earth can now be seen as an assault on ourselves, because we're part of, not separate from, the larger web of life.

WHAT'S POSSIBLE FOR THE FUTURE?
The Great Turning

When we look at some of the great social changes in history—like the British leaving India, the end of the Cold War or the dismantling of South African apartheid— there seems to be a pattern: After many years or decades of business as usual, an unpredicted shift occurs, and dramatic results show up in a very short period of time.

When we look deeper, though, we find that the time before such "sudden" shifts was a kind of incubation period of background work, often marked by the blood, sweat, and tears of thousands of ordinary, passionate, unglorified individuals. Many small actions by many people over time create the environment in which a rapid turnabout becomes possible. This means that *any day can be the tipping point,* and every conscious individual can play a role in bringing it about.

"Can we change? Absolutely. Our species changes when it has to. And that's where I find hope in the despair of the situation.... There's a drivenness in our species and a creativity ... a capacity to let go, to start over, to forgive ... that we've barely begun to tap."

—**Matthew Fox**, *Episcopal priest, founder, Wisdom University*

Clearly something *is* happening on Earth. Virtually everywhere people are creating solutions that form the puzzle pieces of a sustainable, just, and fulfilling world. Are these random, isolated occurrences, or are they signs of a powerful emerging global phenomenon?

Environmentalist and author Paul Hawken, who wrote the Foreword to this book, has concluded that the largest social movement in history is arising—but it doesn't yet know itself as a movement.

"There is another superpower here on Earth that is an unnamed movement... far different and bigger and more unique than anything we have ever seen... non-violent ...grassroots...no central ideology.... The very word 'movement' is too small to describe it.

"No one started this, no one is in charge of it... it is global, classless, unquenchable and tireless... arising spontaneously from different economic sectors, cultures, regions and cohorts...growing and spreading worldwide.... It has many roots, but primarily the origins are indigenous culture and the environment and social justice movements.

"This is humanity's immune response to resist and heal political disease, economic infection, and ecological corruption caused by ideologies. This is fundamentally a civil rights movement, a human rights movement; this is a democracy movement; it is the coming world."

—**Paul Hawken**, *author, environmentalist, entrepreneur*

This movement is accompanied, in many of its adherents, by a resurgence of interest in personal spiritual growth—but the spiritual orientation based on the "new story" is an inclusive spirituality, a spirituality that transcends the strict boundaries of traditional formal religions.

Once you become aware of this movement, you will see it everywhere.

WHERE DO WE GO FROM HERE?
Getting Engaged, Taking a Personal Stand

What this remarkable time in history is asking of each of us is to discover and act upon *what is uniquely ours to do*, while working with others who share our vision. If you're alive today, you have a role to play.

The power of one individual acting on his or her vision and passion cannot be overemphasized. The Green Belt movement (*greenbeltmovement.org*), which plants trees to preserve African watersheds, is a perfect example. What began with founder Wangari Maathai working with a few women on Kenyan hillsides has grown to impact the soil, the climate, and local economies all across Africa. More than 600 community groups have planted over 50 million trees, and the movement is now spreading to other regions of the world. Maathai became the first African woman to receive the Nobel Peace Prize, in 2004.

We have discovered Bigfoot. He is us.

Coming Together in Community

My great-great-grandchildren ask me, in dreams
"What did you do while the planet was plundered?
What did you do when the Earth was unraveling?
Surely you did something when the seasons started failing,
as the mammals, reptiles, birds were all dying.
Did you fill the streets with protest when democracy was
stolen?
What did you do once you knew?"

—*Drew Dellinger*, poet

The story of the monarch butterfly, from evolutionary biologist Elisabet Sahtouris, offers a powerful metaphor for the transformation our world is undergoing.

A caterpillar consumes hundreds of times its own weight before it enters the cocoon stage. Then, so-called imaginal cells[11] begin to emerge. Initially, the caterpillar's immune system sees these new cells as threats, and kills them off. In time, the imaginal cells begin to find one another. When enough of them cluster, they form imaginal buds that can resist attack.

Even though they are not in the majority, the buds become the genetic directors of the organism's future. Other cells continue to dissolve and feed the growing structures of the butterfly, a creature of great beauty that travels great distances catalyzing life.

Imagine that we, who are working to heal the web of life in the context of the new story of interconnectedness, are the planet's imaginal cells. Discounted by the mainstream, our ideas are often "killed off" by the still-powerful old "immune" systems. But as we find each other and collaborate, we are gaining influence. Eventually, we will guide our civilization onto a course that affirms all life on the living Earth.

Another metaphor for this is that we are part of the Earth's immune system (as opposed to the industrial civilization's immune system). We are here to "hospice" the old, resistant, dying structures, and "midwife" the new, Earth-honoring systems and ways of being.

We must remember that imaginal cells only gain power in community. Our individual actions are essential, but only within the larger context of "all my relations."

"There's an old African proverb that says: 'If you want to go quickly, go alone—if you want to go far, go together.' We have to go far, quickly! So we have to have a change in consciousness, a change in commitment, a new sense of urgency, a new appreciation for the privilege that we have of undertaking this challenge."

*—**Al Gore**, former US Vice President, author of* An Inconvenient Truth

In this profound transitional time, in which humans have the opportunity to live extraordinarily meaningful lives, we must embrace the crises in which we're swimming and see that the possibilities are greater than the magnitude of the predicament.

Our work is to contribute to the creation of a global movement of engaged people. As global citizens, we must be informed and compassionate. We must be able to see, embrace, and create new possibilities for the future. And we must be continually learning and growing to ensure that our actions will be as effective as possible in bringing needed change. We invite you to stay in a state of what modern dance pioneer Martha Graham called "blessed unrest"—a vibrant, active state of agitation, empathy, and vision, fueling creativity and committed action.

Three Facets of "The Great Turning"

"Reminded that 'fear is excitement that has forgotten to breathe,' we can see distress as opportunity, the seedbed of the future. Then, as countless men and women are doing in this time of Great Turning, we join hands in learning how the world self-heals."

— Joanna Macy

Joanna Macy popularized The Great Turning as "the transition from an industrial growth society to a life-sustaining society," with "three simultaneous and mutually reinforcing dimensions":

- **Holding Actions:** the front-line, direct actions to stop or limit the immediate damage, as well as all political, legal, and legislative work;

- **Systems Change:** solutions that address structural causes of the crises and offer alternative models;

- **Shift in Consciousness:** a profound transformation of our perception of reality, our values, attitudes and goals.

As we engage "the essential adventure of our time," Macy cautions that there is no guarantee that we will make this transition in time for the survival of civilization or even of complex life systems. It is in the not knowing how the future will turn out that we will find our passion, which will inform and sustain us in striving to bring forth The Great Turning.

"The environmental crisis is an outward manifestation of a crisis of mind and spirit. There could be no greater misconception of its meaning than to believe it to be concerned only with endangered wildlife, human-made ugliness, and pollution. These are part of it, but, more importantly, the crisis is concerned with the kind of creatures we are and what we must become in order to survive."

—Lynton K. Caldwell

EXPLORE & ENGAGE

Each chapter of the *Sourcebook* is followed by an "E&E" section, for you to **explore & engage** with the rich and often challenging material, and connect your daily life choices with your impact on the world. Through reflection, discussions, and experiential activities, it offers many diverse ways to engage. Get inspired, explore issues and community as you take action for planetary change.

Types of activities are indicated by icons. For the sake of variety, personal preference, and learning styles, we offer five primary activity types. Choose ones that interest or challenge you:

⑦ **Discussion questions** invite conversation. For those using the *Sourcebook* on your own, we recommend journaling on the question, then finding a discussion partner.

♥ **Experiential exercises** include role-plays, debates, and field trips. These activities offer new perspectives, direct experience of an issue or solution, and more emotional involvement.

➡ **Action items** encourage specific proactive ways to contribute on an individual, group, or community level.

♀ **Contemplation themes** invite sitting meditatively with a question or topic and tapping into deeper intuition and feelings.

📖 **Research opportunities** involve the use of outside sources such as the internet, books, videos, and other expert sources to broaden understanding of a topic.

You will also see this icon, used within the text (indicating a link rather than activity type):

👁 **View videos, websites and webpages**; the links to the suggested items to see are all found here: *swcoalition.org/explore-links*. If you find that a link is no longer valid, please let us know at *info@swcoalition.org*.

Chapter One: Setting the Context

EXPLORE & ENGAGE

Neil Palmer (CIAT)

The Pachamama Alliance offers the *Awakening the Dreamer, Changing the Dream Symposium* (ATD) for the purpose of **bringing forth an environmentally sustainable, spiritually fulfilling, and socially just human presence on this planet.** The Symposium is presented by volunteers around the world.

Four essential questions structure and shape the Symposium experience as well as the Setting the Context chapter of the Sourcebook: **Where are we? How did we get here? What's possible for the future? Where do we go from here?** These questions lead to a "big picture" understanding with interconnections that are unpacked in the chapters that follow. This big picture context empowers us in our work solving the social and economic issues we face as humanity, and in healing the fragile fabric of life on our precious planet.

1. **Consider the name "Awakening the Dreamer, Changing the Dream." How is it relevant to your own vision of a world that works for all?** (p.8)

   ⑫ How are dreams relevant to bringing visions into being? What visions have you dreamed of, or dream of now?

   ⑨ Reflect on the Joanna Macy quote (p.9). Identify an issue in our culture's history where we "woke up from a dream" and it led to a profound shift in society. *(Examples: ending slavery and the right of women to vote.)*

   ⑨ Read Drew Dellinger's poem (p.14) aloud slowly, or watch his powerful video. ◉ What is your emotional response to the poem?

   ⑫ What poets, writers, songwriters, or artists have sparked your engagement or inspiration? Share with the group.

All over the sky,

a sacred voice is calling you.

—Black Elk

2. **The *Sourcebook* is meant to educate and inspire, using the purpose of the ATD Symposium.**

   ⑨ On a scale of 1-10, how aligned do you feel with this purpose?

   ⑫ Discuss the relationship between the three aspects of this purpose— is it possible to deal with major issues without considering all three? How do you define spirituality?

3. **"Where Are We?"** (p.9) **is a status report on the current health of our planet.**

   ⑫ Focusing on the three aspects of the ATD purpose, discuss facts and challenges of each that are particularly of concern to you. Then share what touches your emotions.

   ⑨ Brian Swimme suggests, "We haven't deepened our hearts in a way that would make possible the grief that is wanting to be felt." Look inside to identify the types of challenges or tragedies that shut you down. Where might your heart be closed? Where is your heart deeply open? What are some ways you might deepen your heart?

4. **"How Did We Get Here?"** (p.11) **examines the trance the modern world appears to be living in. Contemplating other worldviews and our own unexamined assumptions can awaken fresh perspectives and possibilities.**

   ⑫ What unexamined assumptions do you recognize that you have had, which have significantly limited your experience? *(See bulleted examples, p.11)* Make a list of eight such assumptions and what they have cost you. Share in group or with partner.

➡ Take a walk outside and deeply experience each tree, plant, or animal as one of your relations. What shifted or was different when you looked with "new eyes and heart"?

> We have to abandon the conceit that isolated personal actions are going to solve this crisis.
>
> *—Al Gore*

5. Many of our core beliefs and unexamined assumptions are so much a part of our individual and collective identity that they may be difficult to see.

💡 Reflect on: "People take actions appropriate to how they see the world." Recall a time when you shifted a belief or assumption which had important ramifications. How does this relate to the current world situation?

❓ Describe an incident where you acted on an assumption and got yourself in a real pickle, only to realize later that what you had assumed to be true was inaccurate.

6. It is easy to fall into the habit of assuming that our world—and the way we view it—is an objective reality. Look at the following and consider alternative perspectives.

View one or both of the following TED Talks:

- Bill Strickland makes change with a slide show ◉
- Majora Carter on greening the ghetto ◉

❓ How did watching the video expand your view of social and environmental justice? Discuss.

💡 What unexamined attitudes and prejudices allow people from the inner city to be "left for dead"? Look underneath your surface assumptions to see if you can discover your deeper, unexamined assumptions about class/race privilege.

❓ What do you now see as possible from viewing this video?

7. "What's Possible For the Future?" (p.13) reminds us that change can seem difficult or impossible and that history is full of surprises and awakenings that once seemed totally improbable.

❓ Share what you (personally), others you know (community), and groups you have heard of (nationally or globally) are doing now to create more environmental sustainability, social justice, and spiritual fulfillment.

➡ Celebrate each action and movement. Feel the hope engendered.

8. Margaret Mead said: "Never doubt that a small group of thoughtful, committed citizens can change the world; indeed, it's the only thing that ever has."

💡 Think of three examples in history when this happened. Can you imagine how a small group of people could impact a specific major issue in your community?

💡 Who has inspired you? What role(s) did that person/group play (role model, rule breaker, trendsetter, whistleblower, inventor, philosopher, architect)?

❓ Identify a pivotal experience in your life or the life of someone you know. Share the story of how this transformed your or their life.

> We must recognize that we are one human family and one Earth Community with a common destiny. We need to create an exciting, inspiring and joyful image of the future.
>
> —*Jim Embry*

9. "Where Do We Go From Here?" (p.14) focuses on stepping into action and taking a stand about what is important to you.

📖 Leadership receives a great deal of attention. Watch the *Leadership Lessons* video about the crucial role of early followers in creating a movement. 👁

💡 Identify the gap between your current level of being and acting sustainably, justly, and in fulfillment and where you would like to be.

💡 Does either metaphor—the butterfly or the immune system (p.14)—appeal to you to enlist others in taking action to make a difference? Do you have an approach you like better? Share in your group.

➡ If there is an *Awakening the Dreamer, Changing the Dream Symposium* happening near you, sign up for it and invite one other person to join you. 👁

➡ Take a stand by declaring aloud to your group something that you will do/practice that relates to your reason for being on the earth at this time. Consider deepening or enlarging something you are already doing.

💡 What steps could you take to enhance your leadership skills in your field of endeavor, or to build your effectiveness and capacity in what you do?

> Enlightened leadership is spiritual if we understand spirituality not as some kind of religious dogma or ideology but as the domain of awareness where we experience values like truth, goodness, beauty, love and compassion.
>
> —*Deepak Chopra*

Part Two: ENVIRONMENT
HEALING THE WEB OF LIFE

Our culture is based on a principle that directs us to constantly think about the welfare of seven generations into the future.

— Iroquois Confederacy

CLIMATE CHANGE
Life in a Warming World

The human race may one day deeply regret its slow response to the challenge of climate change. We are already facing shifts in the climate and landscape of our planet, from floods and rising sea levels to droughts and melting glaciers. Scientists tell us even more severe and irreversible changes await us in the future unless we take swift and decisive action.[1]

So what's stopping us from reversing our course toward climate catastrophe? In the United States, where just 5% of the world's population creates 17% of the greenhouse gases that cause global warming, divisive politics, entrenched economic interests, and a lack of engagement by citizens have historically stalled meaningful efforts to combat climate change.

There is some evidence that the tide of inaction might be turning. Devastating climate events, like Hurricane Sandy, which pummeled the US mid-Atlantic in 2012, the devastating droughts of 2011-12, and the record-breaking wildfires that scorched the western US in 2012 and 2013, are opening the public's eyes to the impacts of climate change. Meanwhile, in 2013 President Barack Obama announced a climate action plan that aims to reduce greenhouse gas emissions and prepare for the effects of climate change both at home and abroad.

The best weapon in the battle against climate change, however, is a well-informed citizenry that demands action and accountability. Read on to learn more about the very real threats facing our environment and steps we can take to fight them.

"It is not the strongest of the species that survives, nor the most intelligent, but the one most responsive to change."

—**Charles Darwin**,
19th-century English naturalist

Understanding the Greenhouse Effect

A fortuitous mix of gases that are naturally present in the Earth's atmosphere keeps the upper part of our ecosystem in equilibrium and the planet habitable by trapping heat in the thin blanket of air that surrounds us. These gases absorb and re-emit radiation from the sun.

The interplay of natural forces that results in climate warming is called the greenhouse effect. The substances now popularized as greenhouse gases (GHGs) include carbon dioxide (CO_2), methane, nitrous oxide, ozone, and various chlorofluorocarbons (CFCs), including aerosols and gases used in refrigeration and air conditioning. One consequence of increased GHGs is a growth in atmospheric water vapor in response to rising carbon dioxide concentrations. This in turn greatly heightens the warming effect of man-made CO_2 emissions.

The greenhouse effect has been conducive to life on Earth, and for millennia the amount of the main GHG, carbon dioxide, remained fairly constant at about 280 parts per million (ppm). But since the Industrial Revolution, beginning around 1750, the quantity of CO_2 and other gases released into the air by human activity has greatly increased. These emissions—from industry, power plants, motor vehicles and aircraft, among other sources—remain in the atmosphere for many decades, so they will affect the climate far into the future. The rate and volume of these emissions now also exceeds the Earth's capacity to absorb and re-use these gases.

Today's level of CO_2 is 400 ppm—a 40% rise from pre-industrial times in only a fraction of the time humans have been on Earth. Our oceans, forests, and soil act as natural "carbon sinks"—reservoirs for storing carbon. But when the amount of atmospheric GHGs exceeds the carbon sinks' capacity to absorb them, excess heat is retained in the atmosphere.

Women sorting plastics for melting—outskirts of Guangzhou, China

When this happens, air, ocean, and land temperatures rise and global climate change occurs.

The Verdict Is In: *We* Are the Source of Global Warming

Multiple factors cause the climate to change over millennia. Some are natural; some are human-caused. Experts now generally agree that recent increases in global temperatures result primarily from higher levels of heat-trapping gases in the atmosphere. Scientists have also demonstrated that the primary human source of such gases—about 80 percent—is fossil fuel burning. The second human source of GHGs is deforestation and other land use changes.[2]

Meanwhile, livestock farming produces GHGs that are more harmful than carbon dioxide, generating 65% of the world's nitrous oxide (296 times as warming as carbon dioxide), as well as 37% of methane emissions (23 times as warming as carbon dioxide). Methane is largely produced by the digestive system of the animals and enters the atmosphere from their manure and urine. According to the US Environmental Protection Agency (EPA), cow-calf operations in the beef industry are the largest livestock emitters of methane.[3] EPA also suggests that improved manure management and changes in livestock diet could help reduce these emissions.

One-third of all the raw materials and fossil fuels

Big Bend Power Station near Apollo Beach, Florida

used in the US go to raising animals for food[4]—and thus, indirectly, to producing the country's GHG emissions.

According to scientific models, global temperatures could rise by 1.1° to 6.4°C (2.0° to 11.5°F) by 2100.[5] Some scientists fear that we may reach this point as soon as 2050. If we continue on our current trajectory of rapid fossil-fuel growth, over the next century the perfect storm of population growth, resource depletion, and climate change will have catastrophic results.

As of June of 2013, one of the warmest Junes on record, the planet has experienced a streak of warmer-than-average months that has lasted 28 years, according to the National Oceanic and Atmospheric Administration. At the same time, unusually wet conditions in 2013 drenched the Midwestern and eastern US[6] while torrential rains brought flooding in India that killed almost 6,000 people. Meanwhile, about 80 percent of the western US experienced drought conditions.

Among the serious concerns prompted by climate change is the fact that global ecosystems have feedback loops and tipping points, not all of which we fully understand. It is, however, now clear that several of these phenomena are self-sustaining, amplifying cycles. For example, melting ice and glaciers, melting tundra and other methane sources, and increasing ocean saturation with CO_2 lead to increases in atmospheric CO_2.

Once a tipping point in one of these systems is reached, climate feedback mechanisms rapidly speed warming. These processes are nonlinear and largely unpredictable. For all we know, tipping points may be long past or just around the corner. We are unable to accurately specify precisely what quantity of GHGs will be dangerous, though many scientists agree it is much lower than commonly assumed and that we have already set the stage for dramatic and long-lasting changes. In addition to increased global air and water temperatures, these changes include increased and often dramatic variability in local temperatures and precipitation patterns, effects that are already being felt worldwide.

Given the time lag between cause and effect with

regard to global warming, even if we ended all emissions tomorrow, additional warming will still occur thanks to the momentum built into the Earth's intricate climate system. The oceans, for example, have yet to come into equilibrium with the extra heat-trapping capacity of the atmosphere. As the oceans continue to warm, so will the land around them. The velocity and extent of recent changes suggest that climate models are much more conservative than nature itself.

Addressing Climate Change: Urgent and Imperative Government Action Needed Now

The US bears considerable historical responsibility for the problem of climate change, given its contributions to global greenhouse gas emission. It also has the capacity for action on policy fronts. But the lack of leadership by the US, which is second only to China in annual CO_2 emissions, has hampered global progress in addressing the issue.

In the European Union, where per capita emissions are half that of the United States, policymakers and citizens have made much stronger efforts toward reaching the EU's goals for reducing total greenhouse gas emissions. (The EU has committed to reduce GHG emissions by 80% by 2050). And persuading big polluting nations like India and China to curb emissions without a binding US commitment to do the same is unlikely.

Setting GHG emissions reductions goals and reduction commitment time-frames is the focus of ongoing international climate treaties. The most well-known is the 1997 Kyoto Protocol, which set binding

Carbon Tax

To slow the juggernaut of climate change, we as a global civilization must reduce the amount of carbon dioxide released into the atmosphere. Perhaps the most promising potential remedy is the carbon tax—forcing top polluters to pay for the amount of carbon they emit.

The first carbon tax appeared in Finland in 1990. More than 15 countries have since implemented various versions—including some of the most populous countries in the world, such as China, India, Japan, Brazil, the United Kingdom, Australia, and South Africa. Regional carbon taxes have also appeared in some places, such as Quebec, Canada, and some states in the US.

There is no set rule for which sectors of the economy are taxed, nor is there a standard tax rate. Australia, for example, imposed in 2012 a $23 charge per ton of carbon produced. This will shift to a trading scheme in 2015, in which the government will set a limit for total carbon emissions each year. Then companies that pollute more will be able to purchase emission rights from companies that emit less carbon. Meanwhile, to save money, polluters will seek cheaper methods of production that release less carbon into the atmosphere.

The carbon tax is not without controversy. One reason is that consumers will ultimately pay for it, since corporations will pass on the cost of the tax to their customers in the form of increased prices. Furthermore, lower-income populations will likely shoulder proportionately more of these costs since they statistically invest in more carbon-intensive products than wealthier people. To help ease these burdens, the tax is often accompanied by a dividend returned to citizens paying the higher prices.

How carbon tax revenue will be used is also a source of controversy. And some opponents of the tax claim it will lead to "carbon leakage," in which companies skirt the cost by outsourcing carbon-intensive production methods overseas to avoid the financial hit.

There are still many kinks to work out, but if properly implemented, the carbon tax could be a promising and overall cost-effective method of reducing carbon emissions in the foreseeable future.

To learn more about the carbon tax and how you can support it, visit the Citizens Climate Lobby at http://citizensclimatelobby.org/about-us/carbon-tax-and-rebate-faq/

— Beth Vanden Heuvel,
Tri Marine International

Terminology for the 21st Century[a]

Global Warming: As more greenhouse gases are released into the atmosphere from the burning of fossil fuels than are trapped by ocean, forest, and soil carbon sinks, more heat is trapped in the atmosphere, and the average global atmospheric temperature increases, a condition known as global warming.

Climate Change/Global Warming: Climate change results directly or indirectly from human activity that changes the composition of the global atmosphere; this is in addition to natural climate variability. The more popular term "global warming" recognizes that global temperatures overall have been increasing since the Industrial Revolution. Some people have now also begun to refer to the phenomenon of "climate disruption."

Intergovernmental Panel on Climate Change (IPCC): A large, global scientific body tasked to evaluate the risk of climate change caused by human activity.

Fossil Fuels: These fuel sources, composed of hydrocarbons, are derived from fossilized remains of plants and animals found within the top layer of the Earth's crust. Examples include coal, petroleum, and methane. Fossil fuels are considered non-renewable resources because they take millions of years to form, and fossil fuel reserves are being depleted far faster than new ones are being formed.

Carbon Dioxide (CO_2): The main greenhouse gas released into the atmosphere largely through the combustion of fossil fuels. Atmospheric concentrations of CO_2 are estimated to be at their highest level in at least 800,000 years.

Methane (CH_4): This chemical compound is the principal component of natural gas. Burning methane in the presence of oxygen produces CO_2 and water. The relative abundance of methane and its clean burning process makes it an attractive fuel. However, methane's high global-warming potential makes its use less than ideal: Over time, a methane emission has 23 times the impact of a CO_2 emission of the same mass.

Carbon Credit: A financial instrument aimed at reducing greenhouse gas emissions. One carbon credit represents the reduction of one ton of carbon dioxide. Carbon credits are awarded to countries or groups that have reduced their greenhouse gases below their emission quota. Carbon credits can be bought and sold in the international market.

Carbon Footprint: The calculation of combined carbon emissions that a specific activity produces.

Ecological Footprint: An accounting system that measures the total biocapacity for a given region, and compares it against the quantity of resources people use (and pollution people cause) within that same region, thus showing whether there is a positive or negative "resource account."

emissions reduction goals for six primary greenhouse gases. But while the US signed the agreement, it has never ratified it, and this has not encouraged other nations to abide by it. GHG emissions continue to rise worldwide every year, with fossil fuel burning (oil, coal and natural gas) a major contributor.

Go Climate-Neutral: How One Group Did It

Many organizations, public, private, large and small —as well as individuals—are attempting to neutralize their carbon footprints by both reducing overall energy use and emissions and by purchasing carbon credits.

The Center for Biological Diversity (CBD; biologicaldiversity.org), for example, chose to become climate-neutral with a program designed to explicitly track and then maximally reduce GHGs. They also purchased carbon offsets equal to the amount of all past emissions since the organization's inception in 1989. These offsets, which support forest conservation in Madagascar, produce many direct additional benefits to biodiversity and local communities. CBD used the Corporate Greenhouse Gas Accounting and Reporting Standard developed by the World Resources Institute and others to create an inventory of GHG sources including all electricity and heating fuel used in offices, and all automobile and airplane travel on CBD business.

They then calculated average emissions per staff member and multiplied this by the number of staff members in previous years to obtain an estimate of total emissions since 1989. This total is relatively small, about 480 tons of CO_2. After extensively researching carbon offset purchasing options, CBD chose to purchase 500 tons of CO_2 credits in the Makira Forest Conservation Project, Madagascar.

The island of Madagascar, off the east coast of Africa, is one of the world's biodiversity hotspots. The Makira Forest Conservation Project, a joint project of Conservation International, the Wildlife Conservation Society, and the government of Madagascar, will mitigate 9.5 million tons of CO_2 emissions over the next 30 years through forest conservation, including replanting of cleared areas.

WHAT YOU CAN DO

Increasing awareness of the greenhouse gas consequences of our energy use, travel, food, and other choices is the first step toward reducing our own emissions. The average American generates approximately 17 tons of CO2 annually, but this number can be drastically reduced with simple changes, many of which will also save you money. There are many resources available to further reduce your emissions. See Part Seven: Getting Personal for footprint information.

Start with the simple acts of energy conservation suggested in these pages. Electricity, natural gas, propane, and all forms of energy are as valuable as water and must not be wasted. There are environmental costs to their production, transport, and combustion. Be aware and conserve.

Think about your personal carbon footprint on the Earth: How much generated "juice" do you burn during your day? How much "embodied energy" goes into each product you buy—in other words, how much energy did it take to create it and bring it to market?

How much do you drive, walk, cycle or use public transportation in a given week? The Center for Sustainable Economy is one of many organizations that offers a free questionnaire to calculate your Ecological Footprint. The Center's quiz, however, goes beyond measuring just your carbon footprint; it also includes the footprints of food, housing, and goods and services.

Be active in the public dialogue, and talk about this problem to your friends. Help teach others that cutting or reducing emissions of GHGs is a change we all must support now. Once you have reduced your emissions as much as possible, you can go further by supporting organizations working for policy change and by purchasing offsets. Lend your vocal political and financial support to organizations that are serious, committed advocates for policies involving mandatory reductions in GHG emissions.[7]

Urge local, state, and federal government bodies to make appropriate legislation a priority. Tell your representatives at federal, state, and local government levels the following:

■ Act now to limit potential damage from climate change rather than waiting and having to take more costly reactive measures in the future. Timely action could ease the coming impacts of hotter weather, rising sea levels, and bigger storms.

- Adopt federal policies that establish mandatory limits on GHG emissions. Adhere to international agreements.

- Harness the power of markets to drive innovation and protect the climate.

- Don't make carbon-intensive investments in developing countries.

- Climate protection in developing countries must be supportive of economic and social development; foster technical cooperation programs.

THE OCEANS
Deep Problems on the Water Planet

"The frog does not drink up the pond in which s/he lives."
—oral tradition, Teton Sioux

Vast as they are, covering three-quarters of the globe, the world's oceans are not infinite. We are pushing up against the oceans' limits by overfishing, as well as by polluting and dumping waste—including non-biodegradable plastics—into waterways and seas.

Current problems of ocean acidification, rising temperatures and sea levels, and threats to fisheries all stem from cumulative abuse. Only a serious culture-wide reduction in both pollutants generated and CO_2 emitted will stem this rising tide of destruction.

All life on Earth is connected to the oceans. They are our part of our life-support systems, on which the global food web and water cycles depend. Yet human activities are collectively and rapidly degrading the health of the world's oceans.[8]

Global Warming

As major carbon sinks, or absorbers of CO_2 and other greenhouse gases, the oceans are ultimately limited. The oceans' average surface temperature is rising, contributing to weather changes, higher sea levels, current shifts, coastal erosion, and altered fish habitat and migration routes. The oceans have absorbed about half of the CO_2 we've emitted in the last 200 years as it accumulates in the atmosphere.[9]

The ocean is also becoming more acidic, which decreases its capacity to absorb CO_2. Carbon dioxide reacts in the oceans to form carbonic acid, which has increased the water's acidity by 30% in the last two centuries and could increase another 120% by the end of this century.[10]

ARE YOU IN THE KNOW ABOUT 350?

Renowned American environmental activist and writer Bill McKibben has built a global-scale climate change movement, focused on the significance of the number 350. The 350.org campaign (350.org) aims to involve everyone in realizing our collective effect on global warming.

According to McKibben, "350 is the most important number in the world"—the number determined to be the safe upper limit for carbon dioxide measured in parts per million in our atmosphere. If we can make this number known across the planet, that mere fact will exert some real pressure on negotiators. We need people to understand that 350 marks either success or failure for climate negotiations."

With CO_2 levels already exceeding 350 parts per million, McKibben is traveling the world to awaken people to the dire threat of global warming and create a powerful and unified call to action.

This disruption of the oceans' pH balance also threatens marine life, which is unable to adapt so quickly.

Ocean acidification depletes seawater of compounds that organisms need to build shells and skeletons, impairing the calcium-building capacity of coral-forming polyps, crabs, sea urchins, oysters, plankton, and other marine creatures.

A study of reef fish showed that ocean acidification may also alter blood chemistry, decreasing the fishes' ability to sense and avoid predators. Increased blood acidity can also decrease its capacity to hold oxygen, which can affect metabolic health, physical activity, reproduction, and immune system responses. Since the oceans have never experienced a rate of acidification like that currently occurring, scientists do not yet know what the effects will ultimately be.[11]

In one of the many feedback loops we are witnessing in this time of rapid environmental change, rising ocean temperatures are interfering with established ocean currents that move vital marine nutrients upward from deep regions. Without these nutrients in abundance, plankton, which form the basis of the marine food web, will not thrive. In addition to providing essential nutrition to marine species throughout the world's oceans, abundant plankton actually help store CO_2 on the ocean floor when they die and decompose.

Overfishing

The booming human population and its growing appetite for seafood is pushing many ocean species toward extinction. Global seafood consumption has tripled since the 1950s and fish stocks are already collapsing worldwide, largely due to unsustainable fishing practices.

Scientists project that at today's rates of withdrawal, all currently fished species of wild seafood could collapse (experiencing 90% depletion) by 2050.[12] This will not only affect the entire food web, it will decimate the livelihoods and food security of hundreds of millions of people around the world.

Commercial fishing has impacted the oceans in many catastrophic ways. In addition to depleting fish stocks and disrupting the marine food web, commercial fishing has diminished water quality throughout the world's oceans, destroyed habitat, harassed and displaced wildlife, and in many places altered entire marine ecosystems.

And then there is bycatch, the wasteful and unintentional capture of species other than those targeted by a specific fishery. Commercial fishing creates millions of tons of discarded catch annually, including not just fish species but turtles, marine mammals, and seabirds.[13]

Pollution

Waste carried in the world's rivers and streams ends up in coastal waterways and oceans. This includes sewage, industrial waste, and agricultural runoff. Fish absorb industrial pollutants, including hazardous chemicals and heavy metals like mercury that end up in streams and oceans. When we eat such fish, we take in these toxins.

The oceans' ability to absorb hazardous pollutants without adverse impacts to the entire marine food web is declining, but the volume and diversity of contaminants is not. Among the industrial pollutants affecting marine waters worldwide are excess nutrients that have created what are called "dead zones," places where oxygen is so severely depleted it has

Official U.S. Navy Page

Chief Yeoman Ken Vinoya, center, helps gather trash at the Misawa Fish Port, Japan. Air base service members and family helped remove several tons of refuse.

Sustainability For Ocean Health

Go Vegan or Vegetarian. Our diet contributes more to climate change—and the warming and acidification of the oceans—than any other personal cause. We can get the protein and nutrients we need from a plant-based diet, and enhance the health of our world and our body. Get started with delicious options: features.peta.org/VegetarianStarterKit.

Choose Sustainable Fish or Farmed Seafood. If you do eat fish, do so with care—for information on ocean-friendly seafood, consult the recommendations of an organization such as Seafood Choices Alliance at seafoodchoices.org.

Buy Local Products. Support local farmers and fishers. Eating local enhances your community's economy and our global ecosystem. Find out what stores stock local produce and fish (by asking!). See theoceanproject.org/action/involve.php.

Be Trash-Conscious. If you can't recycle, be knowledgeable about what you throw away. For instance, flushing non-biodegradable products can damage the sewage treatment process and end up littering beaches and waters. For other tips on safe trash disposal, visit epa.gov/recyclecity/ or obviously.com/recycle.

Be Considerate of Ocean Wildlife. Our trash can damage or kill ocean wildlife. Never dispose of fishing line or nets in the water. Don't release helium balloons outside. Minimize or reject the use of non-biodegradable plastics such as Styrofoam. Cut open plastic six-pack rings that can entangle ocean life and put them into plastics recycling, ensuring they'll never end up in aquatic environments where they can entangle ocean life.

Reduce Household Toxics. By using phosphate-free detergents and non-toxic cleaning products, you can ensure a healthier ocean and a cleaner overall environment. For more ideas on reducing pollution, see epa.gov/airquality/peg_caa/reduce.html.

Reduce Run-Off. Avoid contributing to non-point source pollution. Don't dispose of hazardous substances down household or storm drains. Use soap sparingly if you must wash your car. Don't use toxic chemicals on your lawn. Scoop pet waste — an estimated 15 tons flows into ocean waters every day. Other ways to reduce your run-off can be found at epa.gov/owow/nps/whatudo.html.

Support or Volunteer on Behalf of Ocean Health. Find a local nonprofit organization working to save the oceans and ocean life, and get involved. For international volunteer opportunities, see oceanicsociety.org.

— from the San Francisco-based Oceanic Society

killed basic components of ocean life.

In the Mississippi Delta, for example, excess nutrients including nitrogen and phosphorus from fertilizer runoff have created enormous dead zones, miles and miles that are now uninhabitable for fish and shellfish.

These nutrients come from large livestock operations and other farms as well as septic systems. Once in the water, the nutrients promote explosive blooms of tiny plants known as phytoplankton, which die and sink to the bottom where they are eaten by bacteria that use up the water's oxygen. This oxygen starvation makes it difficult for fish, oysters, sea grass, and other life to survive.[15] There are now at least 200 of these zones around the world, and they are increasing rapidly.

More than 1.2 trillion gallons of sewage (including human waste, excreted pharmaceuticals, detergents, and household chemicals) and polluted storm water are discharged into American waters annually. In eastern Europe, about 60% of the wastewater discharged into the Caspian Sea is untreated. In Latin America and the Caribbean, the figure is close to 80%. In large parts of Africa and the Indo-Pacific region, it's as high as 90%.[16]

WHAT WE CAN DO

- ■ Support the use of fishing equipment that limits bycatch.

- ■ Limit consumption of seafood in general, and learn which species concentrate toxins.

- ■ Use as little plastic as possible.

- ■ Urge the US Congress to increase government protection of fish stocks. Existing regulations and controls on overfishing must be enforced, primarily targeting those fisheries with the highest rate of bycatch.

- ■ Join a beach cleanup. The Ocean Conservancy (*oceanconservancy.org*) organizes shoreline cleanups each fall. To date, 6.2 million volunteers in International Coastal Cleanups have removed more than 144 million pounds of debris from coasts in over 150 nations.

- ■ Expand your knowledge about marine mammals, their health, and their ocean environment, at the Marine Mammal Center (marinemammalcenter.org).

CORAL REEFS
Colorful "Rainforests of the Sea"

Degraded for decades by toxic run-off, bleached coral reefs are now viewed as early indicators of global warming.

Coral reefs are among Earth's most diverse, exquisite, and fragile ecosystems, essential to the web of life. They have been vanishing at alarming rates for the last 40 years, mainly from damage caused by run-off of agricultural chemicals and waste, invasive species, fishing equipment and overfishing.

Now the rapid pace of environmental change threatens to overwhelm the reef species' ability to survive. Coral reefs appear particularly vulnerable to even the most modest climate-change scenarios, as they are unable to adjust to rapid changes in temperature and ocean acidity. We may be approaching a tipping point that will wipe out corals in entire bioregions.

The World Resources Institute estimates that 75% of the world's coral reefs are at risk due to climate change and localized human activities. Within the next 40 years, many of the remaining reefs may disappear if current emissions trends continue.[17] One-third of reef-building corals are threatened.

The demise of coral reefs affects the entire ocean ecosystem—a quarter of all marine fish species reside in the reefs, according to The Nature Conservancy. The loss of coral reefs is not only devastating to the entire ocean but harms humans as well. Coral reefs provide ecosystem services, such as fish for food, coastal protection, and vital habitat for a diversity of marine life, to over 500 million people.[18]

Marine biologists are now evaluating the remaining healthy coral reefs to determine common characteristics that may provide vital clues on how to best stop continued reef loss. Meanwhile, more than 60 species of coral in US waters are under consideration for legal protection under the Endangered Species Act, which may be crucial to their survival.

Coral reef ecosystem at Palmyra Atoll National Wildlife Refuge

Efforts are underway internationally, by governments and non-governmental organizations (NGOs), to protect coral reefs all around the world.[19]

WHAT WE CAN DO

■ The area of coral reefs under protection needs to be increased globally from the current level of 15% to 30%. Within these protected regions there need to be clear areas where human activities are significantly limited so that already-stressed marine species can recover.[20]

■ Don't touch delicate corals with swim fins. Learn to enjoy snorkeling without touching or breaking the corals, or keep a safe distance.

■ Don't buy coral jewelry, and inform those selling it about the threats to coral.[2]

■ Support organizations that protect corals—by volunteering, donating, or subscribing to one of their newsletters. Organizations making a difference include the Nature Conservancy Institute, the Coral Reef Alliance, the World Wide Fund for Nature, and the National Parks Conservation Association.

FRESH WATER
Using Water Wisely

Nearly 97% of all the world's water is salty or otherwise undrinkable. Another 2% is locked in ice caps and glaciers. Only 1% of the earth's available water can be used without desalinization for agricultural, residential, manufacturing, community, and personal needs. Today, much of that is polluted.

We all know that most life on Earth is impossible without water, so why would we pollute and waste this priceless substance? Bodies of water were long assumed to have an endless capacity to absorb trash and waste of all kinds (including sewage, agricultural effluent, toxic chemicals and other industrial debris). However, rivers, lakes, and oceans worldwide have taken on so much contamination that their ability to support life has been compromised and in some cases is disappearing fast.

Humans currently use about 50% of the Earth's available fresh water, leaving what's left over for all other species. With the world's population increasing, there isn't enough fresh water to waste another drop.

With global warming parching parts of the planet, this will only get worse.

"What you people call your natural resources our people call our relatives."

—**Oren Lyons**, *faith keeper of the Onondaga*

Climate change will exacerbate water shortages. The UN Intergovernmental Panel on Climate Change (IPCC) has predicted that the melting of alpine glaciers and the evaporation of snow cover will accelerate during the 21st century. As a result, many regions will likely experience a decline in fresh water resources that could redirect seasonal water flows and reduce hydropower potential.[22] Water security in a warming world will require major improvements in water-use efficiency—especially in the agricultural and industrial sectors—and in techniques such as rainwater harvesting and groundwater management and use.

Water: A Basic Right Not Yet Available to Everyone

Ensuring access to safe water for all people has long been a humanitarian mission. As part of its Millennium Development Goals (MDG), the United Nations declared its aim to reduce by half the proportion of the population without sustainable access to safe drinking water and basic sanitation by 2015.

According to the 2012 MDG report, although we have succeeded in halving the proportion of the population without sustainable access to drinking water five years ahead of schedule, 783 million people still lack access to reliably clean drinking water. It is estimated that this number will only be reduced to 605 million by 2015.[23]

Meanwhile, more than 2.5 billion people in developing regions remain without access to healthy sanitation facilities. In terms of both water and sanitation, there are still great disparities in access between the wealthy and the poor, as well as between urban and rural areas. How do we ensure that all people will have access to this vital, limited resource? Conservation and wise usage is the only answer.

Historically, governments have been responsible for public water systems, but with the high cost of development and upkeep, along with the urgent need to extend safe drinking water and sanitation systems to the billions worldwide who lack such access, the private sector has entered the picture. The idea is that the public sector is failing to deliver; and the private sector, which is presumed to be more efficient and cost-effective than governments, can pick up the slack.

Institutions like the World Bank and Inter-American Development Bank, as well as private foundations and corporations, are now engaged in developing infrastructure and programs required for municipal drinking water and sanitation systems. There are risks involved, however, to the communities to be served by these water systems.

Transnational corporations are vying for control

True Thirst[b]

40% of the world's population carries their water from wells.

50% of deaths in India can be attributed to poor water quality.

According to the Washington Post, just one flush of a toilet in the West uses more water than most Africans have to perform an entire day's washing, cleaning, cooking, and drinking.

By 2050, more than 523 million people in Africa will not have access to clean water, and famine will be even more rampant as the arid landscape increases.

Forty of the 50 countries on the critical list for water scarcity are located in the Middle East and in north and sub-Saharan Africa.

60% of the world's population lives in Asia, which has only 36% of the Earth's renewable fresh water.

90% of developing nations' wastewater is discharged, untreated, into local waterways.

Each year between 1.8 and 3.5 million Americans are sickened by tainted water.

of freshwater resources as it is very lucrative—in fact, water is being called "the oil of the 21st century." And Climate change will exacerbate this in major ways.

In countries with heavy debt loads and desirable natural resources, privatization can foster corruption and result in higher water utility rates, inadequate customer service, and a loss of local control and accountability, as many privatization efforts have shown. Guidelines have been developed by governmental and non-governmental organizations to help ensure that such private-sector investments result in equitable access and fairly priced services for the communities served.

Concerned organizations are working to protect universal access to safe and affordable drinking water by keeping it in public hands.[24] The social and environmental impacts of water privatization have caused waves of protest as communities all over the world have organized, and in some cases shed blood, to regain control of their water resources. This growing social movement stands in firm opposition to the privatization of our most essential natural resource.

Children in Sindh, Pakistan, play at a village community water pump

The Folly Of Bottled Water

Consumers waste billions of dollars a year on billions of gallons of bottled water. Most of these purchases are unnecessary, create trash, and unwittingly finance the corporate takeover of water supplies. Bottled water can cost up to a thousand times more than tap water and funnels the profits from the sale of water, a public resource, to private companies.[25]

As much as 40% of bottled water comes from a municipal tap.[26]

Production, transportation, and disposal of bottled water consume large quantities of water and energy. You can actually conserve water (and spare yourself the potentially hazardous leaching of plastic chemicals into the liquid) by switching from the bottle to the tap.

Bottled Water Consumption In The United States[c]

1978: 415 million gallons.
2001: 5.4 billion gallons.
2012: 9.67 billion gallons

National Geographic estimates that more than 85 million plastic water bottles are used every three minutes.

WHAT WE CAN DO

■ Stop buying bottled water.

■ Rainwater harvesting can take place anywhere there is a roof. You can gather rainwater in do-it-yourself systems (such as plastic barrels) or commercial systems (for irrigation and livestock).

■ Download and share the Smart Water Guide from Food and Water Watch (foodandwaterwatch.org), filled with facts and helpful tips.

■ Host a movie screening of FLOW, Blue Gold, Chasing Ice, or The Water Front, all powerful documentaries sure to get the message across.

- Curb your own water use. Calculate your water footprint using an online water calculator: h2oconserve.org (part of foodandwaterwatch.org). Some local water utilities can help you reduce water use and storm water runoff so contact your local water company to see if it has such a program.

FORESTS

Home to countless creatures, forests are the lungs of the Earth, arbiters of weather patterns, major storehouses of carbon, and our original cathedrals.

But forests have also long been valued and exploited for timber products, leading to the loss of the great majority of global primary forest ecosystems. According to one estimate, stands of century-old forest now account for only 7% of forest cover in the US.[27]

But lately, trees are being considered somewhat differently—as more than sources of timber and paper—because of their ability to extract carbon from the atmosphere and sequester it in their biomass. Climate change may be the catalyst that prompts trees to possess greater economic value as standing forests than as lumber.

Global financial institutions are even beginning to work out a process for giving carbon credits to businesses that leave trees standing and storing carbon. This also spares the air the effects of pollution that results from burning cut wood—a major contributor to global warming.

Scientists agree that the world's rainforests are, as natural "carbon sinks," the best natural defense against climate change. For example, Indonesian old-growth rainforests store almost 750 tons of carbon dioxide per acre—the equivalent of 620 flights between New York and London. When cleared, rainforests release that carbon into the atmosphere, furthering global warming rather than curbing it.[28]

Halting new deforestation is as powerful a way to combat warming as closing the world's coal-burning plants. But until now, there has been no financial reward for keeping the trees standing. That's what may change: A growing number of experts say that cash payments for preserving valuable standing primary forests are the only way to end tropical forest destruction and "provide a game-changing strategy in efforts to limit global warming."[29]

Carbon Offsets

A "carbon offset" program allows individuals or organizations to compensate for their greenhouse gas emissions by supporting efforts that absorb CO_2 (like forest planting) or reduce greenhouse gas emissions. One emerging and controversial version—Reducing Emissions from Deforestation and forest Degradation, or REDD—seeks to enable industrialized nations and wealthy corporations to offset their greenhouse gas emissions by paying developing nations to protect their forests and/or replant new ones.

However, not all offsets are equal. Poorly designed programs to pay for forest conservation can end up financially rewarding the very people who are destroying them. Verifying the efficacy of a carbon offset effort is a key issue. A major criterion in evaluating carbon offsets is that GHGs can be removed from the atmosphere in ways that would not have occurred otherwise, such as replanting denuded areas.

The Plight of the Tropical Rainforests

Home to more biodiversity than any ecosystem on Earth, the magnificent tropical rainforests are falling at a rate of 100,000 acres per day. That's an area larger than the state of West Virginia.[30]

One reason for this destruction is the rapid proliferation of palm oil and soy plantations. Largely owned by US agribusinesses, these plantations and the rapid expansion of industrial agriculture constitute the fastest-growing threats to the world's great tropical forests in the Amazon, Indonesia, Malaysia, and Papua New Guinea.

Spurred in part by the growing demand for biofuels, companies like US agribusiness giants Archer Daniels Midland, Bunge, and Cargill are establishing soy and palm oil operations in some of the planet's most biodiverse and virginal forests.

Every blade of grass has its angel that bends over and whispers, "Grow! Grow!"

—The Talmud

Palm oil, found in food products, soaps, and cosmetics, could well be the most widely traded vegetable oil in the world today. It is the controversial link between many processed foods, climate change, and disappearing rainforests.

Demand for palm oil has more than doubled in the last decade as worldwide processed food consumption soars. Now that palm oil is also being used for biofuel, the problems are magnified.[31] Farmers are expanding existing palm oil plantations by burning primary forests to plant more palm trees. Nearly all palm oil imported to the US originates in Indonesia.

Soybean production has risen 10-fold over the past 50 years and is expected to double again by 2050,[32] according to a 2014 World Wildlife Fund report. As a primary component in animal feed, soybean demand is largely being driven by the increasing global consumption of meat. For this reason, as well as increasing biofuels production, soybean prices are on the rise—spurring more investment in clearing forests for soy plantations. To learn more about the pitfalls and promise of biofuels, check out Part Three: Energy.

"Biofuels began with a great dream: making fuel from oil or plant waste. But when agribusinesses got involved, the dream went bad."

—*Rainforest Action Network*

WHAT WE CAN DO

■ We can avoid palm oil in products, or at the very least, seek out responsible sources; read the labels carefully! Avoid products derived from recently cleared land and other unsustainable agribusiness practices.

■ Encourage companies that harvest, process and sell palm oil to improve their techniques and sourcing to be part of the solution.

■ Reduce meat consumption and source your meat responsibly.

■ Ask decision-makers in the corporate and political arenas to prioritize proven solutions that halt the expansion of carbon-intensive industries. Policies and investments that support mass transit, bike transit, and plug-in vehicles recharged by a green grid are efficient, cost-effective means to reduce greenhouse gas emissions as well as our dependence on oil.

■ Wood and paper buyers should research the origin of the products they buy. Look for the Forest Stewardship Council (FSC; *fsc.org*) logo as a tool to promote environmentally, socially, and economically responsible management of the world's forests. Learn more about certification standards at these links: credibleforestcertification.org and buygoodwood.com.

Creative Commons/Neil Palmer/CIAT

Aerial view of the Amazon Rainforest, near Manaus, the capital of the Brazilian state of Amazonas

BIODIVERSITY
Every Species Matters

"This we know. All things are connected like the blood that unites one family. Whatever befalls the Earth befalls the sons and daughters of the Earth. Man did not weave the web of life; he is merely a strand in it. Whatever he does to the web, he does to himself."

—attributed to **Chief Seattle**

We are in the midst of the sixth great extinction documented on Earth—the largest since the dinosaur exodus 65 million years ago. According to the Proceedings of the National Academies of Sciences, the current period, known as the Holocene extinction event, may be the greatest such event in the Earth's history and the first due to human actions.

Our hand in the current extinction crisis is undeniable. Every day, we destroy habitat across the globe to satisfy our relentless need to accommodate the exploding human population, obtain essential forest and ocean resources, and convert native ecosystems to arable land (in large part to raise soybeans to feed livestock). Land-use changes, such as deforestation and conversion to cropland, contribute to global warming emissions as well as species loss.

Extinction is accelerating all around us. Although extinction is a natural phenomenon, its normal rate is about one to five species per year. Scientists estimate that presently, dozens of species are going extinct every day.[33] According to biologist E.O. Wilson, the current catastrophic extinction rate is 100 times the normal rate, and it is expected to rise to 1,000 times or higher. And in each of the prior extinction spasms, it took about 10 million years for evolution to regain the amount of biodiversity lost."[34]

If present human consumption trends continue, half of all species of life on Earth could be extinct in less than a century "as a result of habitat destruction, pollution, invasive species, and climate change."[35]

Species loss weakens the web of biodiversity. Given the interrelatedness of all of Earth's species, the extinction numbers are likely to snowball in coming decades if ecosystems continue to unravel. At this point, however, we don't know precisely when the major tipping points will occur or exactly what they'll be.

Meanwhile, according to scientific estimates, every 10 or 20 minutes—the time it may take you to read a few pages—the last individual of a unique species has taken its last breath and has gone extinct forever.[36]

CRIES OF THE WILD

■ 50% of all primates and 100% of great apes are threatened with extinction.

■ Three of the world's eight tiger subspecies became extinct in the past 60 years; the remaining five are endangered.

■ Humans have already driven 20% of all birds to extinction.

■ 12% of mammals, 12% of remaining birds, 31% of reptiles, 30% of amphibians, and 37% of fish are threatened with extinction.

The extinction of species has reached crisis proportions with the collapse of the diversity of life that sustains both ecosystems and human cultures. Extinction is forever.

Major Causes of Extinction

Habitat Loss

Habitat alteration and loss are the primary forces driving species extinction. As human populations increase exponentially, more land is deforested or otherwise altered for housing, farming, livestock, resource extraction, roads, and other uses. Species previously living on that land either move and adapt or die. Efforts to reestablish habitat seldom work and are less stable than the natural systems that evolve over time.

Invasive Species

The second leading cause of extinction is the introduction of invasive species, brought to a new habitat by humans. Not all transplanted species adapt well, but some can take over habitat of indigenous species, often driving them to extinction. Humans can be viewed as by far the most destructive invasive species, since we use 40% of net primary productivity—all organic matter produced by photosynthesis on Earth—and progressively take over indigenous species' habitat, leaving other species to depend on what's left over.[37]

Pesticides and Toxic Pollution

After World War II, the use of new synthetic chemicals as pesticides and herbicides became widespread. These chemicals have been bio-accumulating in plants and animals, including humans, ever since. Rachel Carson's groundbreaking book *Silent Spring* (1962) focused public attention on these chemicals' adverse effects on wildlife.

Other sources of potentially lethal pollution include lawn chemicals, pharmaceuticals, and other industrial chemicals flushed into surface water. Air pollution can also affect wildlife. For example, mercury from coal-fired power plants is found in fish throughout the US.[38,39]

Global Warming

The warming climate is undermining biodiversity by accelerating habitat loss, altering the timing of animal migrations and plant flowerings, and forcing some species toward the poles and to higher altitudes. The speed with which the planet is warming doesn't allow most species to adapt naturally, and condemns many to extinction.

Major alterations to the complex and delicately balanced food web will have significant and often unpredictable impacts. Scientists warn that just 10 more years of our current greenhouse gas pollution trajectory may commit the planet to devastating warming trends, sea-level rise, and species extinction. Many species such as polar bears and penguins are already suffering and severely threatened by the effects of climate change.[40]

Exploitation and Poaching

Overhunting by humans has caused many species to go extinct. Sport hunting of tigers and other large mammals has contributed to the demise of many of the planet's most majestic creatures and top predators. Poaching—for example, the profitable murder of elephants to further the illegal ivory trade—undermines global conservation efforts and must be opposed with all legal means.

Solutions

Many concerned nonprofit organizations assist countries with purchasing and managing refuges for endangered wildlife while simultaneously educating people about the importance of biodiversity, rainforests, and sustainability. See the Resources section for a complete list, and give these groups your support.

In recent years, innovative debt-for-nature swaps have given poorer nations a chance to wipe out some of their international financial burden by protecting vital habitat from degradation or loss.

Meanwhile, as referenced earlier, intact forest ecosystems are being accorded a very different value on the world market—that of carbon credits—in

A Grey Nurse shark (Carcharias taurus), NSW, Australia; shark populations have been decimated by industrial fishing practices and by the shark fin soup market

US Fish and Wildlife Service

Northern Spotted Owl (Strix occidentalis caurina)—near the McKenzie River, Oregon

recognition of their important natural functions, which include carbon storage and climate moderation. This trade-off still needs refinement, but it holds potential for protecting valuable wildlife habitat.

Eco-tourism is becoming a vital component of many national economies, bringing in foreign exchange while minimizing negative impacts on protected wildlife areas. The more travelers pay to visit botanically interesting and/or wildlife-rich areas, the greater the incentive to local communities to value and protect these habitats. However, to maintain natural landscape and wildlife qualities for long-term motives rather than short-term gain, the local community must be a direct economic beneficiary of eco-tourism.

In a groundbreaking example of innovative environmental policy, Ecuador has recently rewritten its Constitution to recognize that nature has the right to "exist, persist, maintain, and regenerate its vital cycles, structure, functions, and its processes in evolution"

Finally, only a concerted international effort to slow greenhouse gas emissions and maintain as much biodiversity as possible will give many species hope of surviving the coming decades, much less centuries.

Recognizing the Rights of Nature

The condition of our planet demonstrates the limits of our environmental laws and other legal systems to protect nature. Our laws treat nature as property for us to manipulate and merely attempt to regulate how much damage we can do to it—rather than recognize the inherent rights of ecosystems and species to exist, thrive and evolve. In the US, dozens of communities have responded by passing ordinances that articulate enforceable rights of nature. These communities and nations such as Ecuador, Bolivia and New Zealand are part of a global movement to transform our relationship with our Earth community and recognize the rights of natural systems to which we as humans belong.

For more information and to join the movement, visit the Global Alliance for the Rights of Nature: theRightsofNature.org.

WASTE, POLLUTION, AND TOXICS

Fouling Our Own Nest

As a species we've created vast amounts of pollution, far more than anyone really wants to know. That's not news if you have read about or experienced tainted food, contaminant-related illness, birth defects, or asthma prompted by airborne particulates.

Indeed, humans are the one species that is poisoning their own—our own—home; and in our case it's the entire planet, so we're taking many species out in the process.

Excess consumption, wasteful practices, our energy sources, and the ongoing manufacture of deadly toxins are at the core of many of our environmental problems. They need to be addressed comprehensively at the industrial, governmental, and personal levels.

While we can lobby our political representatives to enact or enforce laws reining in industrial-scale pollution, most people feel powerless to alleviate these serious problems individually. Yet, there are actions we can each take to reduce carbon emissions, solid waste, cumulative effects from "mildly toxic" home and yard products, and many other apparently small but pernicious forms of pollution.

You might also be inspired to tackle the consumption disease—affluenza—at its core, by reducing your overall rate of consumption and doing all you can to make sure what you do acquire is responsibly produced and sourced, both environmentally and socially. Doing so has a ripple effect: it inspires others around you to do the same. See Part Seven: Getting Personal for details on how to clean up your own act.

Something about this just doesn't feel right.

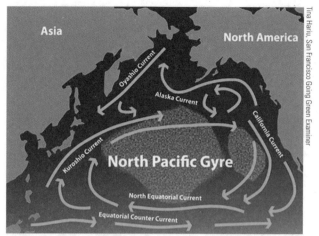

Ocean currents that create the North Pacific Gyre, the world's largest garbage patch

The Plastic Vortices

With all the plastic we produce, use, and toss, it was inevitable that the oceans would eventually teem with windblown and drifting debris. Durable and lightweight, buoyant and persistent, plastic travels over vast distances. The majority of marine debris is plastic. A Texas-sized plastic "vortex" in the North Pacific Ocean has been the subject of study and news reports, but this Great Pacific Garbage Patch is not the only one—the Indian and Atlantic Oceans are also home to seas of garbage.

A plastic vortex, or gyre, gets its name from the dynamics of its formation: Ocean currents and winds create pools or eddies where things can gather. And gather they have, increasing threefold since the 1960s. The Great Pacific Garbage Patch contains at least 4 million tons of plastic litter, including bits of packaging, plastic bags, cigarette lighters, and used diapers. Broken, degraded plastic pieces outweigh surface zooplankton here by a factor of six to one.[41]

The UN Environment Programme estimates that 46,000 pieces of plastic litter are in every square mile of the oceans.[42] The gyre continues to increase due to poor waste management practices on land and sea.

Although plastic products are convenient, the vast majority of plastics don't biodegrade readily, since no naturally occurring organisms can easily break them down. And plastic dust *never* biodegrades.

Many plastics do photodegrade, or break down from the effects of sunlight, into ever tinier bits. Each of these bits of plastic remains in the environment for

Thousands of dead albatross can be found on Midway Island with plastic filling their stomachs

centuries. Most linger near the sea surface where they are mistaken for food by birds, fish, and other marine life. Marine conservation groups estimate that more than a million seabirds and 100,000 mammals and sea turtles die globally each year by getting tangled in or ingesting plastics.[43]

Even the process of photodegradation can take a long time: Estimates include 500 years for a disposable diaper and 450 for a plastic bottle. The more we produce, the more we have to live with—forever.

Project Kaisei (projectkaisei.org), which is surveying the Pacific Garbage Patch and gathering data, is attempting to find ways to clean up and recycle plastic into useful commodities, such as clothing, construction materials, and diesel fuel.

Carrying trash on head—electronic waste in Ghana

POLLUTION: WHAT WE CAN DO

- Agencies at all levels of government need to mandate and enforce emissions reduction targets and pollution limits for cars and trucks, industry and agriculture.

- Recommend to your legislators that subsidies for polluting businesses must be eliminated.

- Ask your legislators to sponsor or vote for environmental tariff legislation: import or export taxes placed on products imported from or being sent to countries with substandard pollution controls. Full implementation of the Basel Convention, designed to prevent the export of hazardous waste from richer to poorer countries, would help achieve this goal. However, even more measures, including extended producer responsibility policies, are needed to prevent export and dumping of waste, particularly in countries where environmental regulations are lax or poorly enforced and where such dumping becomes an example of environmental injustice.

- Lobby for a plastic bag ban in your community. San Francisco became the first US city to ban plastic bags from stores and restaurants.

- Push for laws that require environmentally and socially responsible recycling. Post-consumer material reduces the amount of virgin material consumed.

- Unfortunately, much "recycling" of electronic waste translates to dumping overseas. For information on recycling electronic waste responsibly, visit ban.org or the US Environmental Protection Agency's website (epa.gov/osw/ conserve/materials/ecycling/ index.htm).

- Write to manufacturers and retailers—get them directly involved with plastic disposal and closing the material loop. These companies can have an impact on container and resin makers who in turn can help develop the reprocessing infrastructure by taking back plastic from consumers.

- Advocate for standardized labeling that provides useful consumer information. The "chasing arrows" recycling symbol is ambiguous and misleading. We need different labels for "recycled," "recyclable," and "made of plastic type X."

- Reduce, Reuse, Reduce—then offset and reduce some more.

EXPLORE & ENGAGE

Every part of this earth is sacred; every pine needle, every sandy shore. We are a part of this earth and the earth is a part of us. What befalls the earth befalls all its sons and daughters.

— Chief Seattle

Creative Commons/Devos

Our earth is a spaceship, a remarkable phenomenon of life. Everything we see, breathe, eat, and drink is part of this complex web of interwoven systems. In a very real sense, each of us is intrinsically linked to the whole. The health of this precious blue-green sphere, orbiting a star, affects not only us; the offspring of all living things for generations to come will be greatly impacted by what we do in the next few years.

1. **When climate change is covered in mainstream US media, it is often presented as a confusing and debatable subject.**

   ⑦ How can you identify the most credible climate change websites and determine what organization or individual is behind them?

   📖 Identify three highly credible sites and tell others why each is trustworthy. Share three key findings that are convincing proof of the urgency of taking action.

   📖 Research some of the climate change rebuttal arguments. What points seem to be credible? Identify other points where there may be faulty information or logic.

2. **The concept of a "tipping point" is crucial in environmental discussions.** (p.22)

   ⑦ Define "tipping point" and identify at least two tipping points that concern climate change scientists.

   ⑦ How does the concept of tipping points add to the urgency of taking action on climate change? For further reading, check out this article on tipping points. ◉

3. **The US contributes far more than its fair share (per capita) to global warming, yet has been very slow to take action at the Federal level.**

   ⑦ Name at least 5 factors that have contributed to this delay in action.

   ⑦ What can the US learn from other countries?

- What recent agreements has the global community made to address climate change? (Read: *The Kyoto Protocol*, *The Cancun Agreements*, and one other environmental initiative.) (p.23)

- Form two groups and debate the statement: "The US has played an appropriate role in international agreements."

- Take the role of someone your age in a developing country. Write an email or create a YouTube video that communicates your view of the US's role in climate change, how you feel about it, and how you are impacted.

- Identify the two primary human sources of global warming. (p.22) Identify and discuss the specific sources that contribute most in the US. Do you know the two primary sources within your own community or region?

- List five personal behaviors you do habitually that contribute to global warming. Commit to two specific changes you will make for at least the next month to reduce your own personal contribution to global warming.

4. The greenhouse effect is related to global warming and climate change.

- Define "greenhouse effect" in simple terms. Memorize this definition or have it handy so that you can explain it in one minute or less—regardless of your listener's age or level of education.

- Bill McKibben says, "350 is the most important number in the world." (p.26) What is the significance of this number?

- Look at the 350.org website. ● What rallies, campaigns, information, or messengers anywhere in the world inspire you?

- Share something you learned in a creative way on a social media site

The Earth has a skin and that skin has diseases; one of those diseases is man.

— *Friedrich Nietzsche*

5. Darwin is quoted as saying, "It is not the strongest of the species that survives, nor the most intelligent, but the one most responsive to change." (p.21)

- Read this statement again. How is it powerfully different from the more commonly attributed Darwin phrase "survival of the fittest"?

- "Survival of the fittest" is widely used to support a competitive orientation to life. Discuss how this quote could alter that mindset in business practices.

- Who are some of your (or your culture's) role models? (*Such as Bono,*

Oprah, Mother Teresa, or your aunt who cares for the neighborhood) To what extent does wealth, power, or fame influence who we tend to see as strongest or most "fit"?

6. Oceans cover three-quarters of our earth, and yet many societies seem to take the sea for granted.

⑦ In what ways are people who live far away from the sea nevertheless affected by the health of our oceans?

⑦ Why is ocean acidification sometimes referred to as "global warming's evil twin"? What are the implications of ocean acidification? (p.26)

📖 Witness Sylvia Earle's wish to save our oceans in this moving TED talk. ◉

💡 Write your initial feelings and thoughts about the movie.

⑦ What human blind spots may lead to abuse or neglect of our oceans?

⑦ Define and discuss the implications of:

- Feedback loops
- Bycatch
- Deadzones

Only in the last moment in history has the delusion arisen that people can flourish apart from the rest of the living world.

— E.O. Wilson

7. Coping with disaster, danger, and destruction in the wake of a natural disaster can bring up far-reaching issues that require difficult decisions and ample co-operation.

♥ Group Exercise: Natural Disaster

- Divide your group into four teams.
- Think of a natural disaster in recent history. (*Examples: the 2010 oil spill in the Gulf of Mexico, or the 2011 earthquake and tsunami in Japan*)
- Each team should represent a major player in this disaster. For instance (using the Gulf of Mexico example):

  - A family that depends on fishing for their livelihood
  - An oil company supervisor
  - An organization promoting the health of marsh and ocean
  - A motel owner who relies on tourists who want to fish and enjoy the ocean

- Allow a minute in silence for each group to contemplate the role they represent and to truly adopt that perspective, both emotionally and intellectually.

Creative Commons/Pet R

- Invite each person to speak for two minutes—staying in that role—on his/her perspective. All others listen, suspending judgment.

- Then, as a group, discuss what you have learned and how you might find common ground.

- Conduct a brainstorming session about how nonjudgmental listening might be used to problem-solve issues relevant to habitat destruction. Unleash your creativity and generate as many imaginative ideas and potential solutions as you can. Remember that two guidelines for brainstorming are to exercise no judgment and to ignore "practical" considerations. Have fun with this exercise.

USFWS/Jim Maragos

8. The health of our aquatic ecosystems starts on a personal and community level.

📖 Consider a body of water or waterway near you. Compare its current condition with its overall health 30 years ago. Think in terms of purity, diversity, abundance of life, native plant and animal health, overuse, and run-off. Consider doing personal interviews as well as using the Internet.

➡ Develop a plan to improve the health of this body of water.

➡ What is one actual thing you will do to make a positive impact?

9. Coral reefs are often referred to as "rainforests of the sea" because of their diversity and fragility— and also because they are in peril.

📖 Read this article and discover how this ambitious team is trying to save the rainforests of the sea by growing coral reef nurseries. 👁

❓ List the reasons that the health of worldwide coral reefs is globally crucial.

❓ What are some of the statistics about coral reefs that surprise you? Concern you? Give you hope?

10. Turning on the tap: Examine your own personal water use.

💡 Consider this statistic: Just one flush of a toilet uses more water than most Africans have to perform an entire day's drinking, cleaning, cooking, and washing.

➡ Go to a footprint calculator and calculate your water footprint.

We ourselves feel that what we are doing is just a drop in the ocean. But the ocean would be less because of that missing drop.

— *Mother Teresa*

⇨ Keep a "water journal." For one week, note each time you drink, bathe in, cook with, or use (toilet, garden, laundry) water.

♀ Each time you are about to use water, take a moment to pause and appreciate your easy access to safe, cheap, clean water.

② Discuss what you have learned about your own water use with members of your household and friends. What you will change?

11. Some experts believe that "peak water" presents at least as many challenges as "peak oil."

② Discuss "peak water"—what it means and why it is important.

📖 What are some of the facts/statistics about water (both in the US and globally) that might make a difference to individuals and to policy-making, if they were more widely known?

⇨ With a partner or two, write an email listing these facts/statistics, and send it to at least seven people.

♥ Imagine that your region is considering privatizing water and is having a hearing to discuss the issue. Form small groups (3-5 people) and prepare a 2-5 minute presentation about why it is important that water remain in the public domain. Deliver your presentation to the greater group or a wider audience.

📖 Watch a movie about water (such as *Flow, Blue Gold, or The Water Front*). Discuss and share what you learn.

⇨ Develop a 30-second "elevator speech" about why it is important to use reusable water containers rather than bottled water.

📖 Watch *The Story of Stuff*, which explores the issue of artificial demand. ◉ If you've already seen it, watch it again newly and see what you may have missed.

12. Forests are far more than "resources" or pretty places where animals live.

② Why are forests referred to as "carbon sinks"? Discuss the value of forests. (p.33)

📖 How many acres of tropical rainforest are disappearing a day? (p.33) Calculate how many acres are destroyed each hour, minute, second.

♥ Go outside and time how long it takes you to walk the perimeter of an acre. Determine how quickly that acre of rainforest is being destroyed.

Rainwater harvesting

📖 Watch the video with Julia Butterfly Hill's stand on disposables. 👁

➡ What is the correlation between disappearing rainforests and your use of paper? (*Think: printing, store packaging, bills, coffee cups*) Commit to at least one action you and a buddy will take. Check in once a week to help hold each other accountable.

13. We are currently in an extinction period, known as the *Holocene*, that is greatly impacting the world's biodiversity. Every 10-20 minutes, the last individual of a unique species takes its final breath. (p.35)

- Sit quietly for 10-20 minutes and meditate on the fact that in this time period another species has become extinct.

❓ How has the way we are living brought us to this threshold where 50% of animals are in danger of becoming extinct? How is this extinction different from past ones?

- Brainstorm ways to get the word out about this little-known, under-reported calamity.

📖 Research the killing of dolphins and whales in Japan, Denmark's Faroe Islands, and Greenland. Discuss what you learn and how you want to respond.

> Our culture is based on a principle that directs us to constantly think about the welfare of seven generations into the future.
>
> —*Iroquois Confederacy*

14. Pollution is inarguably one of our greatest environmental problems. We may think, "Pollution is something that other people create."

❓ In what ways do you contribute to pollution? Consider:

- Noise pollution (*use of a cell phone in public, leaf blower, motorcycle*).
- Using disposable items when it is more convenient or you "can't help it."
- Throwing organic material in the trash when it could be composted.

❓ The UN Environment Programme estimates that 46,000 pieces of plastic litter are floating on every square mile of ocean. (p.38) How do the tiny bits of plastic in the gyres and oceans affect marine life? Discuss what can be done about this.

➡ Tour a recycling center.

➡ List things you can reuse instead of throwing away. Be creative. Post it on your refrigerator.

Big Stock Photo

Part Three: ENERGY

Welcome to the dream of renewable energy—a future where our energy consumption doesn't harm the environment.

SMART POWER:
Toward a More Energy-Efficient and Eco-Friendly Future

The fossil-fuel era is coming to an end. Cheap energy has fueled our economic system at the cost of the environment; reversing course requires changing almost everything. We will have to adapt to this change—in both our behavior and our thinking.

Some might call it sacrifice, but we can maintain high standards of living, along with a healthy environment, if we can grow beyond our dependency on fossil fuels. Almost every modern human activity requires coal, oil, or natural gas. This cannot continue without devastating and irreversible environmental damage.

Global energy demands are rising. To feed this hunger, oil and gas companies are tapping difficult-to-reach sources using increasingly polluting extraction methods, such as tar sands mining and hydraulic fracturing (or fracking).

The good news is that an array of options exists. We know how to make energy efficiency and renewable power a reality. The bad news is that it's taking us far too long to identify, agree upon, and implement the best alternatives. We are far behind in our efforts to change our harmful habits. It's time to step up the global push for cleaner energy.

Efforts are underway. The United Nations named 2012 "The International Year of Sustainable Energy for All" and created the Sustainable Energy for All Initiative (se4all.org). Its purpose: to coordinate efforts to (1) globalize access to modern energy, (2) double energy efficiency, and (3) double the world's use of renewable energy by 2030. Meanwhile, communities around the US have launched proactive strategies toward a more sustainable future.

Harnessing renewable energy

According to the Worldwatch Institute (worldwatch.org) and other research centers, solar and wind energy sources are already sufficient to meet the world's demand for energy. What's needed is the commitment and the funding to implement these systems everywhere.

Such estimates indicate that well before mid-century it will be possible to run most national electricity systems with minimal fossil fuels and only 10% of the carbon emissions they produce today.[1] Growing our capacity to put renewable energy to use can also alleviate strife and competition over fossil fuels, diversify national energy supplies (thus enhancing national security) preserve water and air quality, and create "green" jobs—jobs that don't depend on fossil fuels.

Sustainably produced energy also offers low-cost options for reducing CO_2 emissions and fossil-fuel dependency.[2] And yet it occupies only a small fraction—about 10 %—of overall global energy production. Climate change concerns, greater government support, and high oil prices have increased investment in renewable energy, but the level of investment needs to grow much more rapidly to meet even the most minimal targets for climate protection.

ENERGY OPPORTUNITIES TO REDUCE GLOBAL WARMING[a]

The Natural Resources Defense Council (nrdc.org) highlights six major areas of energy-sector opportunities that can significantly reduce global warming:

1. Implement building efficiencies to reduce heating and electricity usage.

2. Increase vehicle efficiency to help cars go farther on less fuel, and design self-sufficient communities to reduce vehicle travel.

3. Employ industrial efficiency to reduce demand for fuel for heat and power.

4. Tap more renewable electricity sources—wind, geothermal, and solar—to supply more of our energy needs.

5. Encourage use of sustainably produced low-carbon transportation fuels, such as biofuels made from crop waste and switchgrass, to replace imported oil.

6. While it's not currently feasible, pursue the development of technologies for carbon capture and storage deep in the Earth, where it can be gradually absorbed. This could make coal-fired power plants a viable component of our energy future.

NRDC is the one of the most effective environmental action group in the US, combining the grassroots power of 1.4 million members and online activists with the courtroom clout and expertise of more than 350 lawyers, scientists and other professionals.

RENEWABLE ENERGY IS A RENEWABLE FUTURE

"Renewable energy is defined as any energy resource that is naturally regenerated over a short time scale and derived directly from the sun (such as thermal, photochemical, and photoelectric), indirectly from the sun (such as wind, hydropower, and photosynthetic energy stored in biomass), or from other natural movements and mechanisms of the environment (such as geo-thermal and tidal energy). Renewable energy does not include energy resources derived from fossil fuels, waste products from fossil sources, or waste products from inorganic sources."

Solar Power

The energy that strikes the Earth from the sun is millions of times greater than our energy needs; its potential for power generation eclipses that of all other renewable energy sources. This, the cleanest and most abundant renewable energy source available, can be actively captured to generate electricity or passively harnessed to heat water and buildings. Yet, solar power accounts for only a fraction of a percent of total electrical output—much less than hydropower or wind energy, which are cheaper to produce.[3]

Why? For starters, sunshine is inconsistent and not abundant everywhere on Earth. For small-scale applications such as hot water and household electricity, solar panels work well (and take up no extra space on roofs), but they can be pricey. For larger applications, mass solar collection requires hefty investment, while transmitting power from where it's generated to where it's needed also presents challenges.

Building the infrastructure to switch to solar would cost more at current prices than continuing to burn fossil fuels, but costs are falling steadily. Some nations, including Italy and India, have already reached grid parity, meaning solar power costs the same as electricity from the grid. Others are close behind.

Meanwhile, a host of governments, utilities, and private companies are developing large-scale solar projects capable of powering tens of thousands of homes each.

Common devices to harness the sun's energy include:

- Solar photovoltaic (PV) cells made primarily of silicon, the same material used in computer semiconductor chips.

- Concentration systems that use mirrors to focus the sun's energy onto a PV cell or heat-transferring fluid to create steam, which spins a turbine and generates electricity. These can be large enough to replace coal-fired power plants.

- Windows, sunrooms, and skylights that allow the sun to passively heat and light buildings.[4]

Wind Power

Humans have been harnessing the wind's power throughout history. Now the same force that our ancestors used to grind grain and power sailing ships holds promise as a key source of cleaner energy.

In fact, the US National Renewable Energy Laboratory estimates that wind power could meet 30% of US energy needs using currently available technology. Meanwhile, the federal lab projects that wind energy's cost will decline further over the next decade, making it the most economically competitive renewable energy technology of them all.[5]

The downside of wind power includes the need to transmit it long distances on an outdated system, susceptibility to damage from storms, and visual obtrusiveness on the landscape. But new design modifications can help mitigate these concerns in the same way that past advancements have decreased noise output and turbine-related bird and bat deaths (although these wildlife impacts continue to be a concern.[6]

Many modern wind farms are already operational in the US, which is second only to China in installed wind capacity. According to the US Department of Energy, there could be more than 250 gigawatts of new wind power added in the US between now and 2030—that's enough to power almost 200 million homes.[7] (Job Alert: There is a huge demand for certified windsmiths, the technicians who keep the wind turbine generators maintained and operating at peak efficiency.)

Using solar technology alone, one-fifth of US electricity needs could be produced on a 1,500-square-kilometer plot of land slightly larger than the city of Phoenix, Arizona. An area covering less than 4% of the Sahara Desert could produce enough solar power to equal global electricity demand.

If photovoltaic panels covered just three-tenths of a percent of the US, a 100-by-100 mile square, they could power the entire country.

The Smart Grid

One obstacle to large-scale renewable energy use is the need to construct and maintain new transmission systems for power produced at remote sites, such as the US desert Southwest for solar and the Dakotas for wind power. Also, wind and solar are variable resources, meaning that the output of turbines and solar plants varies depending on weather conditions. This has created a big need for batteries and other storage devices.

Battery storage technology is progressing rapidly, and will allow a higher percentage of renewable energy to be integrated into the grid.

These are all components of what's come to be known as the "smart grid"—a modernized transmission system that integrates communication and storage technologies to balance the supply and consumption of electricity.[8]

Geothermal Energy

Talk about a power source—the potential of geothermal power, which uses the Earth's heat to generate electricity, defies imagination. The first six miles of the Earth's crust contains 50,000 times more heat energy than all the planet's oil and natural gas resources, according to the US Geological Survey.[9] Plus, unlike variable resources like wind and solar, geothermal resources are considered "base load," meaning they are available 24 hours a day, 365 days a year.

The most common form of geothermal power plant, a flash steam plant, uses water at temperatures greater than 360°F (182°C) that is pumped from underground wells under high pressure and vaporized to power electricity-generating turbines at the surface.

Other benefits of geothermal include reliability and a cost of about 3 cents to 5 cents per kilowatt-hour after installation.[10] Geothermal facilities also occupy a fraction of the space required by wind and solar farms. Although the supply of geothermal energy is virtually unlimited, there are large upfront costs to extract it, not to mention potential environmental impacts.[11]

According to the USGS, "The amount of heat that flows annually from the Earth into the atmosphere is enormous—equivalent to ten times the annual energy consumption of the United States and more than that needed to power all nations of the world, if it could be fully harnessed."[12]

Geothermal borehole house, Iceland

RENEWABLE ENERGY DIY[b]
Turn the sun, wind, or rain into on-demand power at your home.

Assessing the potential:

- If you have a spot on your roof or within a few hundred feet of your house that sees sun from 9 a.m. to 3 p.m. most months of the year, you have excellent solar potential.

- If your site has an annual average wind speed of 12 mph, you are in a good area for a wind turbine.

- If you can divert 40 gallons of water per minute (7.5 seconds to fill a 5-gallon bucket) into a pipe and run it downhill/downstream 100 feet, you can potentially have 400 watts of hydropower, 24 hours a day. If this is a possibility for you, check into local permitting requirements.

Available technologies to use and store renewable resources:

- Photovoltaic modules (PV) use semiconductor technology to convert sunlight into electricity, which can be used in an off-grid system to charge batteries, or in a grid-connected system to directly feed the grid, offsetting part or all of the energy consumed on site.

- Solar water heating systems are a logical way to meet part of your energy demand. Contemporary systems are well designed, reliable, and efficient.

- Wind turbines can charge batteries or directly feed the grid. They are usually designed to make the best use of the available resource, considering variables such as wind speed and turbulence.

- On or off the grid, appropriately sized small hydroelectric generators (aka "micro-hydro" systems) can operate seasonally or year round. They can work 24 hours a day and are capable of generating high-voltage power, which can be sent thousands of feet, enabling access to a more distant water resource. Again, you'll need to find out whether permits are required.

- Inverters change the direct current (DC) power from a battery bank or directly from the renewable source to alternating current (AC), a more common and usable form of electricity.

- Batteries are essential to an off-grid system for storing energy when it's available and accessing it later. However, their production and disposal remain weak links in the alternative energy reality.

Remember, if you are investing now in alternative energy, you may be eligible for federal and state tax credits. Check if your state offers rebates for the installation of grid-connected PV and wind systems.

"This is deeper than a solar panel. I want you to have a clean energy revolution. That's beautiful. But I'm gonna tell you the truth about it. If you stop there, if all you do is have a clean energy revolution, you won't have done anything.

If all we do is take out the dirty power system, the dirty power generation in a system, and just replace it with some clean stuff, put a solar panel on top of this system, but we don't deal with how we're consuming water, we don't deal with how we're treating our other sister-and-brother species, we don't deal with toxins, we don't deal with the way we treat each other...

If that's not a part of this movement, let me tell you what you'll have; you'll have solar-powered bulldozers, solar-powered buzz-saws, and bio-fueled bombers, and we'll be fighting wars over lithium for the batteries instead of oil for the engines, and we'll still have a dead planet.

This movement is deeper than a solar panel. Don't stop there! We're gonna change the whole thing. We're not gonna put a new battery in a broken system. We want a new system! "

—*Van Jones*, author, *The Green Collar Economy*, speaking at the PowerShift '09 conference for youth

These non-fossil fuel options can get a little confusing. Here's the lowdown on existing and up-and-coming energy sources that use plants and waste products to power vehicles.

Biofuels

Biofuels are developed from crops such as corn, soy, and palm oil. While some biofuels are already in widespread use today (most US gasoline is blended with ethanol), their net benefit to the global warming crisis is questionable.

The expansion of biofuel crops through clearing and deforestation releases carbon from the soil and from the loss of vegetation. This makes the world's climate problem worse, because forests, peatlands, and grasslands can actually produce more CO_2 than biofuel production saves, negating any biofuel-related benefit for decades.[13]

Not all biofuels are created equal. First-generation fuels made directly from crops, such as corn and soy, are the least effective due to their agricultural impacts. Second generation fuels, made from the inedible portions of food stocks, specifically cellulose, have a higher potential with 40% lower emissions. Meanwhile, third generation biofuels made from algae have been successfully tested but are yet to become commercially viable.

Creative Commons/ Dominic's Pics

Many designer homes are now being built with very high efficiency, that generate more energy than they consume

In assessing the carbon footprint and sustainability of various biofuels, it's also important to consider how much energy it takes to turn the raw material into usable fuel.

Another complicating factor: prioritizing edible crops for fuel, instead of food, can exacerbate hunger issues. For example, in the US, of all corn planted, 40% will be used to create biofuels rather than food.[14] On the other hand, the uses we put corn to in the name of "food"—like livestock feed, high fructose corn syrup, and other processed food additives—also raises significant health concerns.[15]

BIOFUELS: WHAT WE CAN DO

- Support policies that prevent biofuel production in important sensitive ecosystems.

- Switch subsidies from food-based, first-generation biofuels to second- and third-generation biofuels, which have smaller carbon footprints.

- Incentivize the production of biomethane from sources such as animal manure, landfills, and biomass.

- Accelerate development of cellulosic biofuel technologies and the infrastructure to harvest, transport, and process new crops.

- Incentivize sustainable farming practices, such as low- or no-till agriculture, the planting of cover crops, and the creation of riparian buffer zones.

- Support farmers investing in sustainable fuel crops such as perennial grasses or fast-growing trees.

Source: Worldwatch Institute

Hydrogen

Hydrogen fuel is a promising alternative for transportation since the byproducts of its combustion are simply oxygen and water. Industry is slowly developing sustainable and environmentally neutral methods of producing, storing, and distributing hydrogen fuel for motor vehicles. It is also working to advance fuel cell technologies (the "engines" that convert hydrogen into usable energy).

Widespread use of hydrogen fuel in vehicles will require an extensive investment in new infrastructure. The Natural Resources Defense Council finds that the cheapest, most developed and widely utilized methods of hydrogen production—from natural gas, coal, and oil—are "not necessarily environmentally sustainable." In fact, they may lead to higher emissions. The NRDC predicts "it will take at least two decades before hydrogen can even begin to make a significant contribution to reducing global warming pollution," because clean methods of production, such as from renewable energies (wind, solar, and biomass), are not yet economically viable.[16]

Natural Gas

Like oil, natural gas forms over thousands of years under great pressure as plants and animals decompose in the ground. Natural gas produces less greenhouse gas than petroleum, but its extraction, processing and burning have been shown to contaminate groundwater, pollute the air, and destroy habitats.

The current controversy surrounding natural gas comes primarily from new techniques to extract previously inaccessible methane trapped at shallow depths—"shale gas." Hydraulic fracturing, or fracking, involves pumping underground large amounts of water, chemicals and proppants—industrial sand or other material that props open fissures in rock—to capture trapped gas.

The high pressure exerted through wells can push methane and fracking chemicals into the groundwater, while improperly sealed wells can release methane and other toxics directly into the air. Many of the chemicals used in fracking are undisclosed, while others are known toxins, even carcinogens. The long-term health risks of these chemicals are still unclear.

Other liabilities of fracking include air pollution from hydrocarbon compounds associated with the gas, environmental hazards to communities where industrial sand mining is occurring, and the loss of farmland, wildlife habitat, landscape beauty and tourism.

Preliminary studies show that new and improved regulations can minimize the risks involved in natural gas extraction and keep pollutants at permissible levels. However, the natural gas industry is growing faster than scientists can verify health risks and governments can establish relevant regulations to protect human health and the environment. Already there are about half a million natural gas wells in 30 states, with 391,000 to 805,000 more slated for production by 2035, making shale gas 46% of US domestic production.

At this writing, the Environmental Protection Agency is about to release its final report on the environmental impacts.[17] In the meantime, states are working independently to regulate chemicals and wastewater, decrease emissions, and increase accountability. In states like New York, activists are promoting bans on fracking, maintaining that no regulations can keep this procedure safe (in part because it *uses* huge amounts of precious fresh water—and *threatens* more).

Hydroelectric
How Free Is Its Power?

The benefits of hydropower are legendary: as a "free" energy source and as the gravity-powered use of a renewable resource. Proponents also tout the security of stored water and flood protection that some hydropower dams provide. And some believe that big dams rank among the last century's greatest engineering achievements.

But it has taken decades to recognize the true costs of hydropower. These "great engineering achievements" may also be among our worst acts of environmental destruction. Even small dams can have a large cumulative impact, destroying or blocking access to fish habitat, damaging adjacent habitats including wetlands, and eroding downstream riverbeds and beaches.[18]

Large-scale hydropower projects have often proved even more disastrous. In 2009, China completed construction of the Three Gorges Dam, the largest hydroelectric dam in the world, displacing over a million people, flooding important natural ecosystems, and creating ongoing issues with landslides, earthquakes, and drought.[19]

Furthermore, in destroying natural wetlands and floodplains, dams can exacerbate flooding when it does occur. And global warming-induced weather extremes promise more flooding in the future.

What is new is the discovery that reservoirs, as well as the dam mechanism itself in its churning of the water, are major sources of global warming pollution. Dams and reservoirs release significant amounts of the greenhouse gases methane, nitrous oxide, and carbon dioxide, according to a growing number of scientific sources.

When all the conditions and set-up are right, electricity generation via running water works well

Three Gorges Dam, China

on a small scale in many parts of the world. But as the links between healthy environments (such as a river system) and healthy societies are more widely understood, destructive river infrastructure projects are losing favor in many countries. Unfortunately, they are still being promoted in some places around the world, especially to meet the power demands of rapidly industrializing nations.

Coal
The Dirtiest Fuel

Coal is by far the most polluting and carbon-intensive source of electricity. It currently accounts for more than 30% of US carbon emissions. The coal industry is working hard to make us believe that "clean coal" will help meet the world's energy needs, but its expensive and unproven technology is problematic and questionable. Many experts consider "clean coal" an oxymoron. Al Gore has compared clean coal to a healthy cigarette.

The most promising "clean-coal" technology is called carbon capture and sequestration (CCS). It aspires to capture and bury carbon emissions, thus keeping them out of the atmosphere (for now). As of 2013, there were 72 large-scale integrated CCS projects in the world. While some of these facilities have successfully reached the operation stage, the vast majority are still in construction or planning phases.[20] For this to be an option in a warming world, the coal industry, with government help, must increase investment in research and testing.

Just wondering if you have any extra room in your back yard...

From the mine to the plant, coal, which provides about half our electricity in the US, is our dirtiest energy source. It's a major contributor to global warming and air pollution and a leading cause of asthma and other respiratory and health problems. Coal mining has destroyed ecosystems, particularly where "mountain-top removal" coal mining is practiced. Coal burning also releases toxic mercury into our communities. While coal historically has been relatively inexpensive, more stringent pollution regulations and increased construction costs are rapidly making coal-burning plants more expensive.[21] This will continue to make renewables more competitive.

The coal industry spends millions on public relations and media exposure for its "green" efforts, but like the oil and gas industry, it works simultaneously to block and delay effective federal legislation that aims to cap GHG emissions and require utilities to include more renewable energy sources in their portfolios.

There is a growing trend in the US against coal-fired power plants. In 2012, the EPA proposed an emissions cap for new power plants, putting pressure on coal-fired plants to clean up their act. Over 200 existing coal-fired plants—the equivalent of 31 gigawatts of power—are expected to close by 2017.[22]

Nuclear Power
A Risky Climate Solution

Because safe and healthy power sources like solar and wind exist, we don't have to rely on risky nuclear power.

In 2011, Japan experienced the world's worst nuclear incident in 25 years when a major earthquake and tsunami caused meltdowns at the Fukushima Daichi nuclear power plant. The resulting fallout caused evacuations and air, soil, water and marine contamination. By 2014, contaminated seawater had spread 5,000 miles to the west coast of the United States.

This disaster has made several countries, like Germany and Japan, rethink their nuclear power policies. Germany, the largest economy is Western Europe, has pledged to shut down its remaining nuclear power plants by 2022.[23]

Ten Strikes Against Nuclear Power[c]

1. **Nuclear Waste** will be toxic for humans for more than 100,000 years. We can't securely store all the waste from the plants that exist now. Therefore to scale up is unthinkable.

2. **Nuclear Proliferation:** We can't develop a domestic nuclear energy program without confronting proliferation in other countries.

3. **National Security** is at risk because nuclear reactors (which are not entirely secure) are an attractive target for terrorists.

4. **Accidents**—human error or natural disasters—can wreak just as much havoc. The Chernobyl disaster forced the evacuation of nearly 400,000 people, with thousands poisoned by radiation. The Fukushima disaster also displaced thousands and has poisoned farmland and contaminated fishing grounds.

5. **Cancer risk** for childhood leukemia and other forms of cancer seems to be higher in communities near nuclear plants—even when a plant has an accident-free record.

6. **Not Enough Feasible Sites** exist on Earth for new nuclear facilities, which must be near water for cooling, and safe from droughts, flooding, hurricanes, earthquakes, and other potential triggers for a nuclear accident.

7. **Not Enough Uranium:** Scientists in both the US and the UK have shown that if the current level of nuclear power were expanded to provide all the world's electricity, our uranium would be depleted in fewer than 10 years.

8. **Costs** for nuclear power increase with scale, unlike some types of energy production, e.g., solar power, which experience decreasing costs to scale.

9. **Private Sector Unwilling to Finance:** Due to all of the above, the private sector is largely taking a pass on the financial risks of nuclear power.

10. **No Time:** We have the next ten years to mount a global effort against climate change—not enough time to build enough new nuclear plants.

Reproduced, with some editing for length,
courtesy of Rocky Mountain Institute (rmi.org)

Despite its obvious risks, some people tout nuclear as a potential climate solution, since its CO2 emissions are relatively low (although constructing nuclear power plants and supplying them with fuel carries a large carbon footprint).

Nuclear power generates 13.5% of the world's electricity. There are approximately 435 nuclear plants around the world, with 104 in the United States alone. Though no such plants have been built in the US for more than 30 years, the US is planning at least 13 new ones, with the Department of Energy looking to invest in smaller reactors.[24]

"New nuclear power is so costly that shifting a dollar of spending from nu- clear to efficiency protects the climate several-fold more than shifting a dollar of spending from coal to nuclear. Indeed, under plausible assumptions, spending a dollar on new nuclear power instead of on efficient use of electricity has a worse climate effect than spending that dollar on new coal power!"

— *"Forget Nuclear" by **Amory B. Lovins**, **Imran Sheikh**, and **Alex Markevich***

ENERGY CONSERVATION

Reducing energy use and increasing energy efficiency (i.e. using the least amount of resources to produce the maximum output) are now considered the most economical ways of reducing dependence on fossil fuels.[25] Energy efficiency also offers some of the cheapest options for reducing CO_2 emissions.

Conservation in homes and, on a larger scale, by business, government and industry, including agriculture, makes a huge difference in overall energy consumption and climate pollution.

It's not about sacrifice; it's about being smarter in how we spend our natural capital. According to renowned energy consultant Amory Lovins:

Increasing energy end-use efficiency—technologically providing more desired service per unit of delivered energy consumed—is generally the largest, least expensive, most benign, most quickly deployable, least visible, least understood, and most neglected way to provide energy services. The 46% drop in US energy intensity (primary energy consumption per dollar

The Brower Center in Berkeley, CA is a model of sustainable, mixed-use development (and home of Sustainable World Coalition). Using energy-saving technologies and recycled building materials, the Center has as light a footprint as possible, taking into account the true life-cycle cost of building construction, operation, and maintenance.

of real GDP) during 1975–2005 represented by 2005 an effective energy "source" 2.1 times as big as US oil consumption, 3.4 times net oil imports, 6 times domestic oil output or net oil imports from OPEC countries, and 13 times net imports from Persian Gulf countries.[26]

In California, which developed the first energy efficiency standards for appliances and buildings, per capita electricity consumption has remained level for the past 30 years, while the national average has risen by 40%, according to the US Energy Department—proof that such measures can work.[27]

If the rest of the US followed California's lead, we could forego new coal plants, which would be a huge leap forward in the fight against global warming. In general, cities and states continue to be far ahead of the US government in enacting legislation to reduce GHG emissions.

The Sierra Club and its allies, with their Beyond Coal campaign, have secured the retirement of over 150 coal-burning power plants, representing over 55,000 megawatts and over 1% of the existing coal fleet. Their ambitious goal is to retire one-third of US coal installed capacity by 2015.

Right now over 70 percent of the world population is convinced that something serious has to be done about the dangers facing the planet... Most of humanity wants to know how to make the change. It's one of those tipping-point times where things can change unbelievably fast...

—*Paul H. Ray and Sherry Ruth Anderson,*
authors of "The Cultural Creatives:
How 50 Million People are Changing the World"

Transformation to me means the powerful unleashing of human potential to commit to, care about, and change for a better life. Transformation occurs when people give up their automatic way of being and commit themselves to a different future, recognizing that they can influence the flow of events and thus create new futures — individually and collectively.

—*Monica Sharma, UN Director of Leadership*
and Capacity Development

WHAT WE CAN DO

- Insist on and be part of the shift to clean, renewable energy. Laws as well as individual activities must aim to rein in global warming pollution and transition to a clean energy economy. Shift to renewable energy and energy efficiency in all possible aspects of your life.

- In order to harness the potential of renewable energy, we need a strong Energy Efficiency Resource Standard (EERS) requiring utilities to generate 20% of their electricity from renewable resources by 2020. Such a standard would reduce global warming, create jobs, and save consumers money.

- Climate, energy, and transportation policies must complement one another and aim for similar goals.

- Urge government not to favor giant power plants over distributed solutions; nor to emphasize enlarged supply over efficient use.

- Call your local utility company and sign up for renewable energy. If they don't offer it, ask them why not.

- Make your home energy efficient. This yields dramatic savings in heating and cooling. California building codes have resulted in an energy savings of $30 billion since 1975, more than $2,000 per household. Roll those policies out nationally, and the savings would be immense. Department of Energy Best Practices Guidelines can be found at eere.energy.gov/buildings/building_america.

- Start with caulking and weather-stripping on doorways and windows. Then adjust your thermostat and start saving. Ask your utility company to do a free energy audit of your home to show you how to save even more money.

- Many homes and offices waste enormous amounts of electricity through passive energy use. Turn off electronics and lights, and unplug appliances when not in use.

- Replace incandescent light bulbs with compact fluorescent bulbs, which use about a quarter of the electricity and last ten times as long. LED bulbs use about 75% less energy than incandescent bulbs and can last up to 50 times longer.

- Buy energy-efficient electronics and appliances. Look for the Energy Star label on new appliances or visitenergystar.gov to find the most energy-efficient products.

EXPLORE & ENGAGE

Fcmlaw.com

We know we are running out of fossil fuels. The challenges of weaning ourselves from these energy sources are huge. Opening the creative field to include more than technological solutions invites a perspective that may contribute to a world that works for all.

1. A key concept in understanding our current energy situation is "peak oil."

📖 Explore the way oil companies (*Exxon, BP*) define "peak oil" versus the way it is defined in more neutral publications (*UN reports, Christian Science Monitor*).

- Michael Ruppert offers his view on peak oil in relation to the big Gulf Coast oil spill. 👁

- This *Christian Science Monitor* article provides a neutral view on peak oil. 👁

❓ In simple terms, describe peak oil and several reasons it is critical to understand it. What will life be like in a world without oil? Imagine...and discuss.

➡ Talk with a young person or a group of 8-12 year olds about what "peak oil" means and why it is important to take action.

2. Petroleum is pervasive.

📖 Look at a list of petroleum-based products 👁 and identify at least five products most people wouldn't know are made from petroleum.

➡ Write a blog or social media post sharing what you learned about these five "hidden" petroleum products.

💡 Take a day and look at each item you encounter and consider how petroleum is a component or a part of its manufacture or distribution process.

OR

💡 Walk around your home. Imagine throwing out every object that uses petroleum either in its manufacture or distribution to get to you. What would be left?

➡ Make a personal commitment to cut at least one petroleum product out of your life, for good.

❓ Discuss the notion of "oil addiction." In the field of addiction treatment, clinicians draw a distinction between substance use and substance abuse. Where would you draw the line with petroleum?

3. The fossil fuel era is coming to an end.

📖 Research the total percentage of US and global energy that comes from fossil fuels.

❓ What does this mean for you? Your community? People around the world? And for our planet? Discuss.

• In this TED video, Rob Hopkins discusses the transition to a world without oil. 👁

❓ What cultures currently seem to thrive without fossil fuels? To what extent are they truly petroleum or coal free?

4. The good news is that an array of renewable energy options exists. (pp.46-50)

❓ What does it mean for an energy source to be "renewable"?

📖 Learn more about the variety of options available at the Renewable Energy Institute website, *renewableenergyinstitute.org.* 👁

➡ Make a list of how you, your household, and your community could begin to reduce energy use on a daily basis. Start by putting one of these possibilities into action.

5. Compare fossil fuels with renewable energy sources.

❓ List ten different sources of power or energy. For each one, identify at least two pros and cons. After discussion, rank the ten sources in order from most to least desirable. Consider factors such as availability, impact on climate change and the environment, cost, and the current state of development.

❓ Based on what you have learned, what recommendations would you give to policy makers/CEOs/politicians? Discuss

6. Sun, earth, wind, and water are all viable energy options. (pp.46-50)

Charging stations in San Francisco

♥ Take a field trip to explore renewable energy sources. Go to a wind farm, a hydro- electric plant, a geothermal facility, a solar project, or a wave/tidal system.

➡ Record your trip with video, photos, sketches, and/or notes.

② Report back verbally on what you learned.

➡ Post an account of your trip on the Internet.

➡ Build and use a solar shower or solar oven. Document and share your experience.

♥ Take about an hour to use the Internet and gather information on renewable energy sources. Include futuristic "out of the box" ideas that may seem like science fiction; for example, research pavement made of photo-voltaic solar energy cells ◉ Make a 5-minute pre-sentation to a group about one of the futuristic ideas.

② What is a "smart grid"? How is it crucial to renewable energy? (p.49)

- This government website offers a wealth of information about smart grids ◉

➡ Summarize your overall findings.

7. Imagine yourself as an energy system.

♡ How do you feel when you are "energized" versus when you feel "depleted"?

♡ What are the sources of your energy? (*eating, listening to music*) Where do you spend energy? (*working at a computer, driving*) How sustainable are you? Where are you out of balance?

➡ Contemplate where you are efficient and balanced versus where you are energy-deficient. Commit to making one change to better serve your own personal energy reserves.

8. What does Van Jones mean when he says: "This movement is deeper than a solar panel"? (p.51)

② What needs to change on a global level in order to achieve a more sustainable, just, and spiritually fulfilling world? Share your vision.

➡ Make a list of the top five collective attitude shifts you would like to see in order to create a sustainable world.

♡ Where in your own life are you "creating anew" as opposed to engaging in the old paradigm?

📖 Watch the movie *Kogi* 👁 and discuss how this relates to Van Jones's "We want a new system."

9. On many fronts, other countries are leading the way when it comes to renewable energy.

📖 Which countries use the highest percentage of renewable energy? The lowest per capita use of energy? How is their economy? What relationships do you see between these facts? How does the United States' renewable energy use compare to other countries? 👁

❓ Of the energy used in the US, a very small percentage comes from renewable sources. List some of the reasons that the percentage is low in the US, while the per capita use of energy is so high. What challenges might be unique to the US when compared to other "developed" nations? (*Political? Financial? Technological? Citizen will?*)

10. There are indicators of hope—they are clearly visible when we know what to look for.

📖 Identify some groups, companies, and countries that are making significant progress with renewable energy.

❓ What are you most enthused about in terms of alternative energy development and efficiency strategies? Research this and share in your group.

- Donald Sadoway discusses "The missing link to renewable energy," and his work at MIT to create a liquid-metal large-scale energy storage battery. 👁

- Check out this extraordinary breakthrough regarding photosynthesis in this article. 👁

Some solutions are relatively simple and would provide economic benefits: implementing measures to conserve energy, putting a price on carbon through taxes and cap-and-trade and shifting from fossil fuels to clean and renewable energy sources.

— *David Suzuki*

Part Four: A JUST SOCIETY
A World That Works for Everyone

We are way more powerful when we turn to each other and not on each other, when we celebrate our diversity, focus on our commonality, and together tear down the mighty walls of injustice.

—Cynthia McKinney, Green Party candidate for US Presidency, 2008

What would it take to achieve a truly just and sustainable world? Across the planet, we would need to claim our evolved selves and shed the skin of patriarchal attitudes, economic exploitation, and the Industrial Age mindset. All human institutions would need to be guided by the intent to honor the undeniable fact of our interconnectedness.

"Imagine all the people, living life as one," sang John Lennon in a ballad that still brings tears of hope to many. Gandhi's metaphorical "road," the work of uplifting and uniting humanity, has countless twists and turns, obstacles, and difficult choices along the way. But we can share stories and strategies on the journey, lifting each other toward the Holy Grail of unity amid our miraculous diversity.

This section offers a glimpse into both our collective humanitarian crises and the inspiring efforts to address social inequities and alleviate human suffering. On the "problems" side, we face overwhelming challenges like overpopulation, unspeakable hunger, and rampant human rights violations perpetrated on indigenous peoples, women, immigrants, and children in far too many places around the world.

On the "solutions" side, the work being done all over the world inspires profound hope. In examples ranging from a schoolyard garden for young mothers to the urban Green Jobs Corps, we'll explore United Nations declarations, the work of visionary leaders, and new educational opportunities for all ages.

For more background on sources of social justice issues and promising solutions, see the next chapter, Part Five: Economics.

"Where, after all, do universal human rights begin? In small places, close to home—so close and so small that they cannot be seen on any map of the world. Yet they are the world of the individual person—the neighborhood he or she lives in, the school or college he attends, the home, factory, farm or office where he works."

—Eleanor Roosevelt

GLOBAL CHALLENGES

Hunger

According to the Food and Agriculture Organization, "there are 130 million fewer hungry people today than there were 20 years ago, yet one in eight people still go to bed hungry."[1] While progress has been made toward closing the gap, root causes behind world hunger have yet to be resolved. According to the UN, each year, malnutrition is responsible for almost half of the deaths of children under five.

Hunger is the result of three key factors, according to the World Hunger Education Service (WHES): "the neglect of agriculture relevant to very poor people by governments and international agencies; the worldwide economic crisis; and the significant increase in food prices in the last several years which has been devastating to those with only a few dollars a day to spend."[2]

The UN Millennium Campaign's director says, "since the inception of aid (overseas development assistance) almost 50 years ago, donor countries have given some $2 trillion in aid. And yet over the past year, $18 trillion has been found globally to bail out

banks and other financial institutions. The amount of total aid over the past 49 years represents just 11% of the money found for financial institutions in one year."[3]

"When a poor person dies of hunger, it has not happened because God did not take care of him or her. It has happened because neither you nor I wanted to give that person what he or she needed."

—Mother Teresa

People on the Move

Migration is a complex and shifting reality. A host of factors drives people to migrate. In addition to the tragically familiar fate of refugees from war and genocide, people are increasingly uprooted by natural disasters and environmental and resource pressures.[4]

"Climate refugees" is a new term for those forced to relocate due to ecosystem disruption, such as increased desertification, sea-level rise, water shortages, and resulting health epidemics.

The quest for better social and economic opportunities is another reason for migration. According to a United Nations report on human mobility and development: "For many people around the world, moving away from their home town or village can be the best — sometimes the only — option open to improve their life chances. Migration can be hugely effective in improving the income, education and participation of individuals and families, and enhancing their children's future prospects. But its value is more than that: being able to decide where to live is a key element of human freedom…

"There is no typical profile of migrants around the world. Fruit pickers, nurses, political refugees, construction workers, and computer programmers are all part of the nearly 1 billion people on the move both within their own countries and overseas."[5]

Overpopulation

Human overpopulation engages every concern related to sustainability, from food supply, healthcare, and education to climate change and natural resource economics.

The good news is that our best and most effective strategies to create a more environmentally sustainable and socially equitable world may also curb population growth. Actively improving the lives of people worldwide — by making sure women and girls have access to education, making health care and family planning widely available, and alleviating poverty — is the best way to prevent human population from rapidly increasing.

Today there are more than 7 billion humans on the planet. Consider the historical exponential rate of population growth: In 1804, the world population reached one billion. It took 123 years to double that to two billion, 32 years to reach three billion, and 15 years to reach four billion.[6] If the human population keeps growing at this rate, the total number of people on Earth will top 9 billion by 2050.

This booming growth is exceeding the planet's "carrying capacity"—its ability to re-generate depleted resources and to absorb all the waste we produce. This is largely due to the traditional economic model of consumer capitalism, which emphasizes rapid growth, wealth accumulation, consumerism, and dependence on fossil fuels.

The population question isn't as simple as expanding numbers of people using more resources. The rate of consumption per person, which differs greatly among nations, is also significant.

Population growth constantly pushes the consequences of any level of individual consumption to a higher plateau, and reductions in individual consumption can always be overwhelmed by increases in population. The simple reality is that acting on both, consistently and simultaneously, is the key to long-term environmental sustainability. The sustainability benefits of level or falling human numbers are too powerful to ignore for long...

The magnitude of environmental impacts stems not just from our numbers but also from behaviors we learn from our parents and cultures. Broadly speaking, if population is the number of us, then consumption is the way each of us behaves. In this unequal world, the behavior of a dozen people in one place sometimes has more environmental impact than does that of a few

A girl in primary school, Bandar Abbas, Iran

hundred somewhere else.

— *Robert Engelman*, Scientific American

Energy consumption, of course, also increases both when populations grow and when poor people become richer. Reports *Scientific American*: "[The] one-two punch of population growth followed by consumption growth is presently occurring in the world's two most populous countries -- China (1.34 billion people) and India (1.2 billion). Per capita commercial energy use has been growing so rapidly in both countries ... that if the trends continue unabated, the typical Chinese will out-consume the typical American before 2040, with Indians surpassing Americans by 2080. Population and consumption thus feed on each other's growth to expand humans' environmental footprint exponentially over time."[7]

Stemming population growth will require education, the universal adoption of sustainable lifestyle practices, and policies that support both. Societal progress should be measured by genuine well-being, rather than increased production and consumption.

Women

Social justice is impossible as long as women around the world are denied rights.

According to Hillary Clinton, without women's rights and responsibilities, "many of the goals we claim to pursue in our foreign policy are either unachievable or much harder to achieve."

Supplying aid to poor countries is most economically efficient when applied toward health, education, and microfinance initiatives, especially when such programs focus on women and girls. This is because once females are empowered to participate economically in society, more of the family's money is spent on nutrition, medicine, and housing. Consequently, children are healthier.[8]

Economically, socially, and geopolitically, women have tremendous contributions to make.

According to the United Nations, if women farmers had access to the same resources that male farmers do, there would be 150 million fewer hungry people on the planet.

But in many places they face enormous obstacles, including domestic violence, abuse, lack of education and marginalization. In too many places around the world, women are also subjected to the horrors of mass rape, genital mutilation, sex trafficking, bride burnings, and honor killings.[9]

Providing access to education, economic opportunities, medical care, and family planning services not only empowers women but also gives them choices that reduce population growth.[10]

Educated women typically delay childbearing and have fewer children. According to the International Institute for Applied Systems Analysis in Austria, "women with no schooling have an average of 4.5 children, whereas those with a few years of primary school have just three. Women who complete one or two years of secondary school have an average of 1.9 children, which leads to a decreasing population. With one or two years of college, the average childbearing rate falls to 1.7. And when women enter the workforce, start businesses, inherit assets and otherwise interact with men on an equal footing, their desire for more than a couple of children fades even more dramatically."[11]

Enabling the Global Rise of Female Empowerment

Allow women access to educational opportunities, contraception, family planning and the right to choose how many children to have and when. Educated women will transform a community—or a nation.

Allow women access to economic resources such as bank accounts, personal credit, and microfinance opportunities.

Change patriarchal practices such as, in the event of the husband's death, giving the family's property to the husband's brother. Countries that give aid to poorer nations can insist on changing these kinds of laws as a condition of that aid.

Overall, the goal is to raise women's self-respect and social/economic status by affording them equal participation in their communities.

Providing global access to contraception is about five times cheaper than implementing low-carbon technology to combat climate change, according to a report from Optimum Population Trust and the London School of Economics.[12]

" The transformation of women's roles is the last great impediment to universal progress … So-called women's issues are stability issues, security issues, equity issues."[13]

—*Hillary Clinton, US Secretary of State*

EQUITABLE EARTH PRINCIPLES

The United Nations (UN) was founded in 1945 to maintain peace and security within and among nations, and to promote social progress, better living standards, and human rights around the world. Three of the most fundamental instruments developed by the UN address basic human rights for all people. These are the Universal Declaration of Human Rights (UDHR), the Millennium Development Goals (MDGs), and the UN Declaration on the Rights of Indigenous Peoples.

While many nations have embraced these documents' goals, it will take a multitude of organizations working on many levels to turn their promises into reality. Unfortunately, these goals get little media attention; many people do not even know they exist.

Universal Declaration of Human Rights

"Protecting and empowering the poor must become an urgent rallying cry to honor the spirit, the letter, and the promise of dignity for all," states the Universal Declaration of Human Rights (UDHR). This document was adopted more than 60 years ago to promote equality, justice, fairness, non-discrimination, and dignity for all people across all boundaries everywhere and always.

Still, hundreds of millions of adults and children don't have enough to eat.

The UN says that "poverty is often both a cause and a consequence of human rights violations.... Destitution and exclusion are intertwined with discrimination, unequal access to opportunities, as well as social and cultural stigmatization."

Removing discriminatory practices and policies, along with providing pathways out of poverty, serves to "remove barriers to labor market participation and give women and minorities access to employment."[14]

UN Millennium Development Goals

The MDGs, to be achieved by 2015, are drawn from the Millennium Declaration of 2000. Adopted by 189 leaders from the north and south, MDGs explicitly recognize the interdependence among growth, poverty reduction, and sustainable development. These build on the foundations of democratic governance, the rule of law, respect for human rights, and peace and security.

Wikimedia Commons/Jonathan McIntosh

Slums built on swampland near a garbage dump in East Cipinang, Jakarta, Indonesia

1. Eradicate extreme poverty and hunger
2. Achieve universal primary education
3. Promote gender equality and empower women
4. Reduce child mortality
5. Improve maternal health
6. Combat HIV/AIDS, malaria, and other diseases
7. Ensure environmental stability
8. Develop a global partnership for development

How are we doing with these goals? According to the UN's annual Human Development Report for 2013, the unprecedented growth of developing economies like India and China has lifted hundreds of millions of people out of poverty. The report credits education, social programs, health care, and an increasingly interconnected world for helping to have met the 2015 poverty eradication target, which called for halving the number of people living on less than $1.25 per day.

The report, however, stresses that meeting that goal alone is not enough. More than 1.5 billion people—one out of every five people on the planet—live in poverty. And a failure to take swift global action on climate change and other environmental issues could reverse hard-won gains. By 2050, the number of people in extreme poverty could grow by more than 3 billion as a result of environmental disasters, the report says.[15]

Furthermore, not everyone is benefitting from these economic shifts; the wealth is very unevenly distributed. The UN Permanent Forum on Indigenous Issues said of the MDGs, "Indigenous and tribal peoples are lagging behind other parts of the population in the achievement of the goals in most, if not all, the countries in which they live, and indigenous and tribal women commonly face additional gender-based disadvantages and discrimination."[16]

As with any process that concerns them, "including indigenous peoples in the MDG context requires a culturally sensitive approach, based on respect for and inclusion of indigenous peoples' world-views, perspectives, experiences, and concepts of development."[17]

Indigenous Rights

Although indigenous people only account for 5% of the world's population, they account for over 15% of the world's poor. The United Nations counts a total of 370 million indigenous people in more than 70 countries with 5,000 distinct indigenous cultural identities.

The UN Declaration on the Rights of Indigenous Peoples was overwhelmingly adopted in September 2007, but notably not by Canada, Australia, New Zealand, or the United States. The Declaration covers protection of cultural property and identity, the right to education, employment, health, religion, language and culture, and the right to own land collectively.

It says, "Indigenous peoples have the right to maintain and develop their political, economic and social systems or institutions, to be secure in the enjoyment of their own means of subsistence and development, and to engage freely in all their traditional and other economic activities."[18]

Indigenous peoples and NGOs have invoked other agreements over the decades in order to assert indigenous peoples' rights to land and to being consulted on projects that affect them, reports the MDG Monitor, a UNDP information-sharing project.

"Before the day is over, an Indigenous person will be displaced or killed. Before the month is over, an Indigenous homeland will be clear-cut, strip-mined, or flooded. Before the year is over, dozens of Indigenous languages will vanish forever. Governments and powerful economic interests perpetrate this human and cultural devastation."

— *Cultural Survival, a global partner of indigenous peoples, supporting their autonomy and welfare*

INDIGENOUS SURVIVAL

In 1989, the first binding international convention was held to address the rights of indigenous peoples and to hold governments responsible for protecting these rights.[19]

In 2007, First Peoples Worldwide *(firstpeoples. org)* established Keepers of the Earth to protect the rights of indigenous peoples, including rights related to subsistence hunting and gathering, access to sacred sites, and traditional and cultural practices.

First Peoples' development strategy aims to balance biodiversity protection and sustainable economic development in indigenous territories. Their focus is on culturally appropriate management of the diverse community assets of indigenous cultures, such as land, natural resources, traditional knowledge, culture, etc.

Today, the work of keeping attention and advocacy focused on the rights of indigenous peoples is shared by many exemplary organizations around the world. For 40 years, Cultural Survival has partnered with indigenous peoples to protect their lands, languages, and cultures, educate other communities about their rights, and fight against marginalization, discrimination, exploitation, and abuse.

In collaboration with the Calvert Group, First Peoples created the first social investing screen that protects the rights of indigenous peoples.

Indigenous Support

We can all ally ourselves with efforts to improve opportunities and living conditions for indigenous peoples, so their cultures, wisdom, and traditions are not lost, which would be a loss for us all.

Dishchii' Bikoh' Apache Group from Cibecue, Arizona, enacting the Apache Crown Dance

Acknowledge Sovereignty: Tonya Gonnella Frichner, the chairwoman of Seventh Generation Fund's Board of Directors, says of sovereignty: "It's not something that's given to you—it's an expression of who you are as a nation, as a people. It's an act—not only something you think about, it's something you do…. Most important to me as an individual is that the act of sovereignty bears with it a great deal of responsibility."[20]

Seek Environmental Justice: The rights of indigenous peoples have been systematically violated by oil, gas, timber, and mining industries seeking to exploit resources on tribal property. This has created "unconscionable destruction to traditional territories that have sustained us for time immemorial," according to the Indigenous Environmental Network *(ienearth.org)*. It represents a pervasive and centuries-old pattern of assault on native rights as well as on the Earth.

For example, in Alberta, Canada, multiple First Nations are suffering from large-scale oil extraction from the Athabasca Tar Sands, an area the size of Florida in which crude oil is extracted and sent to the US via pipeline. According to the Indigenous Environmental Network, "Current tar sands development has completely altered the Athabasca delta and watershed landscape, with deforestation, open-pit mining, de-watering of water systems and watersheds, toxic contamination, disruption of habitat and biodiversity and disruption to the indigenous Dene, Cree, and Métis trap-line cultures."[21]

But First Nations' resistance to such environmental and cultural threats is increasingly sophisticated. Joining together strengthens the movement. For example, the Indigenous Environmental Network and the Indigenous Women's Network represent grassroots Native communities across North America. Their partnership gives them great collective strength.

Honor the Earth *(honorearth.org)* works to leverage financial resources, articulate Native issues within a wide political context, and reach a non-Native public to provide support and solidarity. The group views energy policy as one route to environmental justice. Through its Energy Justice Initiative, it supports

Southwest Conservation Corps—crew of Native American workers doing preservation work

renewable energy while working to stop fossil fuel extraction on tribal lands.

The Navajo Green Jobs Coalition was organized "because Navajo citizens are sick of working in dirty jobs that pose serious risks to their health, land, and water."[22]

Founded in 1989, White Earth Land Recovery Project *(welrp.org)* is one of the largest reservation-based nonprofits in the US. Its multi-issue approach addresses root causes of problems faced by residents of the White Earth Reservation, including the loss of land, culture, and self-determination. Located in northern Minnesota, White Earth is the homeland to the Anishinaabeg of the Mississippi Band, also known as the Ojibwe or Chippewa.

"Over the past 20 years ... we have purchased or recovered 1,400 acres of land, begun restoring a traditional food system, expanded a wild-rice export market, and initiated a renewable energy program," said founding director and former Green Party Vice Presidential candidate Winona LaDuke.[23]

Wild rice is at the core of our being.... For us, rice is a source of food and also wisdom. For the globalizers, it is just a commodity to be exploited for profit. The paradigms are at loggerheads.... That philosophical, spiritual, and cultural dialogue needs to be deepened in our own communities, because it's in our hands to determine the future.

—Winona LaDuke,
Native American activist, economist

What You Can Do

■ Read and tell your friends about the Universal Declaration of Human Rights, which affirms that, "Everyone is entitled to all the rights and freedoms set forth in this Declaration, without distinction of any kind, such as race, color, sex, language, religion, political or other opinion, national or social origin, property, birth or other status." Learn more and take action by visiting EveryHumanHasRights.org.

■ Get your organizations, schools, and businesses involved. Hang the Declaration poster on your wall. Spread the word.

■ Participate in actions that support human rights at home and around the world.

■ Support equitable pathways out of poverty.

■ Participate in one of the many Oxfam International campaigns (*oxfam.org*).

■ Join with Human Rights Watch (*hrw.org*) to help defend and protect human rights.

■ Advocate for Human Rights First (*humanrightsfirst.org*) policy change proposals.

■ Support the world's most vulnerable through One.org.

■ Support the Human Rights Defenders Policy Reform program of The Carter Center. (cartercenter.org/peace/human_rights/defenders/index.html)

■ Help make human rights real for people around the world through GlobalRights.org.

■ Stand up for the rights of endangered indigenous peoples with CulturalSurvival.org.

■ Connect with one of 10,000+ organizations protecting human rights on WiserEarth.org, and access their state-of-the-art networking tools

HONORING DIVERSITY

Environmental racism refers to any environmental policy, practice, or action that has a disproportionate negative impact on communities of color. The concept of environmental justice provides a framework for communities of color, as well as low-income communities, to articulate the political, economic, and social assumptions underlying environmental racism and injustice.

Environmental justice principles prioritize public good over corporate profit, cooperation over competition, community and collective action over individualism, and precautionary ("do no harm") approaches over unacceptable risks.

The Occupy Movement: A Need for Systemic Change

The effort to achieve social justice is most often a collective one. In July 2011 the anti-consumerist Canadian magazine *Adbusters* called for a massive peaceful gathering on Wall Street, the heart of New York City's financial district. The article, "#OccupyWallStreet," calls for "Democracy not Corporatocracy."[24]

On September 17, 2011, protesters flocked to Zuccotti Park, near the New York Stock Exchange, to voice their dissatisfaction with "illegal foreclosures, exorbitant student debt and the outsourcing of labor."

By mid-November of 2011 the protesters in Zuccotti Park had been removed, but not before making a lasting impression. The Occupy Movement took root in cities across the US, widely represented by the phrase "We are the 99%." The movement highlighted the importance of freedom of assembly and the impact that people can have on large systems when they work together.

And yet, social justice is also furthered by the decisions we make as individuals. Many environmentalists, committed to sustainable lifestyle practices, nevertheless exclude social justice from their definition of "sustainability." While we need to eat locally, make Fair Trade purchases and practice energy conservation, we cannot achieve a truly sustainable society until we work to ensure that every human has adequate food, water, and shelter, as well as equitable access to basic human rights like education, healthcare, economic opportunity, and justice.

While he was campaigning for President in 2008, Barack Obama said, "America can change. That is the true genius of this nation… It requires all Americans to realize that your dreams do not have to come at the expense of my dreams; that investing in the health, welfare, and education of black and brown and white children will ultimately help all of America prosper."[25]

"The next American revolution will be radically different from the revolutions that have taken place in pre- or non-industrialized countries like Russia, Cuba, China, or Vietnam. As citizens of a nation that had achieved its rapid economic growth and prosperity at the expense of Native Americans, African Americans, Latinos, Asian Americans, and peoples all over the world, our priority has to be correcting the injustices and backwardness of our relationships with one another, with other countries, and with the Earth." [26]

—*Grace Lee Boggs*, founder,
Boggs Center to Nurture Community Leadership

Climate Justice

Climate change is a social justice issue. Across the world the consequences of climate change are impacting poorer communities and nations disproportionately. Issues already plaguing the poorest communities, such as fresh water access and food security, are expected to get significantly worse as a result of climate change.

A year before Hurricane Katrina, the book *African Americans and Climate Change: An Unequal Burden* was published by the Congressional Black Caucus Foundation. It concludes: "When it comes to US environmental conditions, African-Americans are … on the frontline of the likely social, environmental and economic upheavals resulting from climate change, … disproportionately burdened by the health effects of climate change, … less responsible for climate change than other Americans, … [and] will be disproportionately affected by extreme weather events, storms, hurricanes."[27]

Internationally, poor countries are already being hit hardest by climate change. These trends are expected to continue. In Bangladesh, for example, rain-fed crop yields could decrease by 50% by 2020, severely threatening the impoverished nation's food security.[28]

These concerns have given rise to a movement

for climate justice, reflected in the work and mission of many organizations around the world. The Ella Baker Center for Human Rights *(ellabakercenter. org),* founded in 1996, has a vision for "justice in the system, opportunity in our cities, and peace on our streets." Its blog summarizes a report on climate equity that "uncovers what researchers call a 'climate gap' or hidden pattern revealing that poor people and people of color in the United States suffer more from environmental changes than other whiter and wealthier Americans."

The occurrence of heat-related illness and death are much higher in low-income neighborhoods where the percentage of household income spent on water is three times higher. The report also offers climate change adaptation investment solutions that can alleviate pollution, create jobs, and lower the cost burden on communities.[29]

Latino Communities and Immigration

The United States has three times as many immigrants ($40+ million) as the next largest country's immigrant population. Immigration numbers in the US have increased more than 30% since 2000 and are continuing to grow. More than half of all US immigrants come from Latin American countries.[30]

Many Hispanic/Latino populations, especially in North America, are confronted with social and environmental injustice, including an immigration system in dire need of reform.

According to the National Council of La Raza (NCLR), the largest national Hispanic civil rights and advocacy organization in the US, the country needs to establish a fair, humane, and practical immigration system that is responsive to the needs of the economy and that encourages legal behavior. NCLR works through its network of nearly 300 affiliated community-based organizations, conducting policy analysis and advocacy in five key areas: assets/investments, civil rights/immigration, education, employment and labor rights, and health.[31]

For Latinos as well as other non-English speakers, the removal of language and cultural barriers is essential for educating families about healthy lifestyles and for increasing their access to health services.

One strategy to deal with these issues is *promotores de salud* (promoters of health) programs. Also known as community health workers, peer leaders, patient navigators or health advocates, promotores

Creative Commons/Voces de la Frontera

"Day without Latinos" march, Milwaukee, 2007—a push for badly needed immigration reform

play an important role in fostering community-based health education and prevention. They fulfill this role in a culturally and linguistically appropriate way, particularly in communities that have been historically underserved and uninsured.

Participants access meaningful healthcare and support services through community activism, advocacy, information exchange, and art as social justice.[32]

Promotores de Salud are well known throughout Mexico and Central America. Sisters of Color for Education hosts the oldest Promotora de Salud program in Colorado. The program has recently been recognized by the US Centers for Disease Control and Prevention (CDC) as a "best practice" for its effectiveness in Latino communities.

Asian American Communities

One quarter of all US immigrants originate from South and East Asia.[33] From the beginning of US history, Asian-Pacific Americans have faced governmental and institutionalized discrimination. The most abusive example was the brutal internment of approximately 120,000 Japanese Americans during World War II. It was not until 1965 that discriminatory quotas against immigration from Asia were discontinued.

The past 20 years has seen a continual increase in hate crimes against Asian Americans. The level of racial profiling against South Asian Americans in particular has increased tremendously since 9/11.[34]

The overall poverty rate for Asian Americans is 13% (compared to 10% for whites), according to 2010 US Census data. Hmong and Cambodian Americans have poverty rates of 38% and 29%, respectively.

The Asian American Justice Center *(advancingjustice-aajc.org)* promotes the human and civil rights of Asian Americans through a growing network of nearly 100 community-based organizations. AAJC's Reuniting Families Campaign works to secure common sense solutions for our family-based immigration system

Considerable health disparities persist in the Asian-Pacific American community. Removing barriers to accessible, culturally and linguistically appropriate healthcare is a critical goal because of the serious health trends for population.[35]

The Asian Pacific American Heritage Festival, hosted by the Coalition of Asian Pacific Americans in New York, is one of the largest celebrations of Asian Pacific American Heritage Month in the US

Forming an exemplary model of much-needed collaboration among diverse groups focused on environmental health, the Asia Pacific Environmental Network *(apen4ej.org)* and four other groups jointly published *Building Healthy Communities from the Ground Up: Environmental Justice in California.* "Every day," the report says, "communities of color across California face challenges of poverty, toxics and pollution, unsafe and unsustainable work conditions, and a lack of safe, affordable housing and basic goods and services."[36]

PATHWAYS TO AN EQUITABLE FUTURE

"Whatever you love opens its secrets to you."

—*George Washington Carver, renowned agricultural scientist, educator, and humanitarian.*

Green Jobs: Repairing Ecosystems

Restoration—from the top of water catchments to urban centers and down to rising seas—offers an environmental, economic, and social win-win. It repairs and restores the natural capital upon which all life depends, and it also helps provide the sustenance, in the form of "green jobs," that can lift families and entire communities out of poverty. The side benefits are numerous, including enabling energy innovations, and making community gardens and sustainable farms possible.

Repairing ecosystems is labor intensive, as tree planters, cone pickers, wildland firefighters, and forest restorationists know well.

Forest worker cooperatives helped develop the concept of green-collar jobs during the timber boom of the 1970s and '80s. Like other large-scale agricultural sectors, forestry has a history of social and ecological exploitation. These days, however, there's a lot more focus on removing roads (before a flood does), fixing trails, reestablishing fish habitat, and thinning forests as part of a caretaking, natural resource-based economy.

The Society for Ecological Restoration *(ser.org)*

and its Indigenous Peoples' Restoration Network support the integration of traditional ecological knowledge into ecosystem repair and management. SER International has thousands of members in 60 countries and 50 states. Its Global Restoration Network is a hub of information on ecosystems, toolkits on how to repair them, and links to caretakers and funders.

In addition to SER, US state and federal agency programs host a variety of restoration programs, often with the cooperation of multiple, diverse NGOs with a conservation focus.

Green Jobs Initiatives

Taking the Ella Baker Center's Oakland, California-based Green Collar Jobs Campaign to the national level, former Presidential advisor Van Jones cofounded Green For All *(greenforall.org)* in 2007 with Majora Carter, founder of Sustainable South Bronx *(ssbx.org)*.

Green For All's intention is to build a green economy strong enough to lift people out of poverty with green job creation and training and urban energy-efficiency retrofits. Through its website, learning communities, and conference calls, members connect with each other to find out what works, leverage their strengths, and build partnerships. Green For All's website includes tools and resources for removing barriers, diversifying funding, and evaluating programs. The organization's Energy Corps and Green Pathways Out of Poverty programs connect disenfranchised and underrepresented groups with specialized training that will help them to succeed in green economy careers.

Green For All helped ensure that green jobs provisions were incorporated into the American Clean Energy and Security Act of 2009.

The Center for American Progress *(americanprogress.org)* and the International Council for Local Environmental Initiatives *(iclei.org)* are currently urging cities to sign on to the Local Government Green Jobs Pledge, through which more than 1,000 cities around the world share models, training, software, and other resources.

The Corps Network *(nascc.org)*, a modern Civil-

ian Conservation Corps, has 27,000 members ages 16 to 25 in all 50 states that, in conjunction with nearly 300,000 local volunteers, "generate 13.5 million hours of service every year."[37] Participating youth work on public and park lands, help with preparation and recovery from disasters, renovate housing in low income neighborhoods, and provide after-school education programs. Credited with improving employment and earning gains, particularly for young African-American men, Corps members show a one-third drop in arrest rates and a drop in teen pregnancies.

Inspiring Youth Engagement

Since youth have little say in how society is structured, some consider them among the ranks of the disenfranchised, and thus worth considering under the umbrella of social justice.

At the very least, the future calls us to transmit our deepest life-affirming emotions, intentions, and actions to our children and their children. At best, youth could be involved in all domains of endeavor that make communities sustainable. Adults must find creative ways to connect youth of all ages with world-centric values and specific sustainability efforts.

The arts, a vital strategic element in social change work in general, are also an ideal vehicle for youth engagement. The arts involve the whole person, and can easily and profoundly engage the whole community, and the whole story, in ways that can powerfully build community resilience, community sustainability.

Models are sprouting up all over the world. Manchester Craftsmen Guild *(mcgyouthandarts.org)*, located in Pittsburgh and founded by Bill Strickland in 1968, was established to help combat the economic and social devastation experienced by the residents of his predominantly African-American North Side neighborhood. Through educational experiences that employ art and enterprise, students have enhanced and revitalized the economic, physical, and social conditions of their communities.

Based on the Guild's success, a vocational education program for adults called the Bidwell Training Center offers multi-disciplinary courses in photography, fine arts, jazz, greenhouse production of orchids, and much more. Strickland envisions a guild in every US center city.

In addition to the arts, food and the sources of our food supply offer vehicles for youth contributions to both community and sustainability. For example, renowned chef and author Alice Waters initiated the Edible Schoolyard *(edibleschoolyard.org)*—a model for a school-based, hands-on, sustainable food curriculum—at the Martin Luther King, Jr., Middle School in Berkeley, California.

Project Sprout is an organic, student-run garden on the grounds of Monument Mountain Region High School in Great Barrington, Massachusetts. The garden supplies the school's cafeteria with fresh fruits and vegetables, helps feed the hungry in the community, and serves as a living laboratory for students.

Located on Detroit's west side in a distressed neighborhood with many empty lots and dilapidated houses, the Catherine Ferguson Academy *(catherinefergusonacademy.org)* supports middle and high school students who are pregnant or already parents. It offers prenatal and postnatal care for the students and well-baby care for the children in onsite day care. The onsite farm produces both livestock and crops and provides the young women an opportunity to dig deep into their own humanity. The girls and women at the academy have a 99% rate of graduation, with 90% of graduates going on to post-secondary schools, and an 80% rate of no second pregnancy.

Educational experiences that allow children to tap into their natural creative gifts are still too few and far between in our culture. These experiences are often marginalized as electives or extra-curriculars, meaning not part of the "important" core curriculum. Programs such as Dawna Markova's Smart Wired *(smartwired.org)* are dedicated to helping kids and young adults maximize their capabilities and bring out their best in all areas of life, not just academics.

In these ways, our youth, no matter their social circumstance, can become part of the shift to a truly just and sustainable world—a shift that begins in their local communities.

"I am here to speak on behalf of the starving children around the world whose cries go unheard. I am here to speak for the countless animals dying across this planet because they have nowhere left to go... I am not afraid of telling the world how I feel... We [the youth] are not the leaders of tomorrow; we are the leaders of today!"

—Severn Cullis-Suzuki,
UN speech at the Rio Earth Summit, 1992

Photo Courtesy of NatureBridge

NatureBridge, in San Francisco, connects young people to the wonder and science of nature

The Seven Foundations of a Just, Sustainable World[a]

1. ECONOMIC FAIRNESS

A world dedicated to economic fairness would strive to meet each person's basic needs, so that no one would lack food, shelter, clothing, or meaningful work. Strength of character and passion should determine an individual's opportunities rather than the economic circumstances into which they were born. In such a world, everyone would benefit from economic prosperity.

- **Challenges:** Economic inequality, debt crisis and unfair trade, sweatshops and other unjust, hazardous and inequitable working conditions.

- **Goals:** End of global poverty, fair trade in all commerce, ethical economics, regulation of multinational corporate practices to support environmental sustainability and justice and human rights.

2. COMPREHENSIVE PEACE

A world committed to comprehensive peace would shift its creative energies towards cooperation rather than competition for resources, resolving conflict rather than escalating it, seeking justice rather than enacting revenge, and creating peace rather than preparing for war.

- **Challenges:** War and genocide, militarization, unilateralism, culture of violence.

- **Goals:** International cooperation, demilitarization and regulation of weapons sales, nonviolent culture.

3. ECOLOGICAL SUSTAINABILITY

A world committed to ecological sustainability would create a new vision of progress that recognizes that the future of humanity depends upon our ability to live in harmony and balance with our natural world.

- **Challenges:** Resource overconsumption, pollution, global warming, overpopulation.

- **Goals:** Clean energy sources, sustainable resource use, stable population growth, global cooperation.

4. DEEP DEMOCRACY

A world built on deep democracy would empower citizens to participate in shaping their futures every day, not just on election day, provide broad access to quality information and democratize our most powerful institutions.

- **Challenges:** Lack of democracy, money in politics, media control by corporations with vested interests.

- **Goals:** Open and honest politics, democratic media, full civic participation.

5. SOCIAL JUSTICE

A world dedicated to social justice is a place where everyone receives respect and equal access to jobs, education, and health care regardless of race, ethnicity, gender, age, physical or mental abilities, economic background, or sexual orientation.

- **Challenges:** Gender inequality, racism, heterosexism, inadequate health care, prisons based on punishment rather than rehabilitation.

- **Goals:** Equal rights for all (including rights for all living things), universal health care and education.

6. CULTURE OF SIMPLICITY

A culture of simplicity would encourage each person to find meaning and fulfillment by pursuing their true passions, fostering loving relationships, and living authentic, reflective lives rather than by seeking status and material possessions.

- **Challenges:** Advertising overload, commercialization of childhood, hyper-consumerism.

- **Goals:** Reclaimed consciousness, a culture of simplicity.

7. REVITALIZED COMMUNITY

A revitalized community would create a healthy and loving environment for people to celebrate their many shared values while embracing individual differences, and would provide support for each person's physical, emotional, and spiritual needs.

- **Challenges:** Loss of connection, lack of compassion.

- **Goals:** Revolution of caring, smart growth (cities designed to support people interacting with each other), strong local institutions.

The Seven Foundations are excerpted from The Better World Handbook.

EXPLORE & ENGAGE

The vision of a peaceful and equitable world has inspired poets and activists for generations. On the ground, the stories and statistics of injustice may lead us to tears or guns, discouragement or contribution. Yet Einstein's quote, "No problem can be solved from the same level of consciousness that created it" invites us to see and act with new eyes and hearts.

The heart that breaks open can contain the whole Universe.

— *Joanna Macy*

1. **What does the phrase "human rights" mean to you?** (p.62-67)

   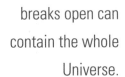 When you think of "human rights," what specific rights come to mind? Where did those concepts originate? Discuss.

   How are children's rights or animal rights related to your idea of human rights?

   Consider two possible worlds at the far ends of a human rights spectrum.

   - Imagine a world where human rights are not valued, where society is shaped by beliefs such as "might makes right." Here, people may be treated with utter contempt and abuse. They have no recourse, and others wonder why they would even think to protest or complain.

   - Now imagine a "global village" where human rights are universally respected. Picture each individual treated with such reverence that the thought of even needing to defend someone's rights seems archaic.

   - Picture yourself living in each of the above scenarios. Identify at least three specific ways that your daily life would be different. What institutions would be lacking or significantly changed?

   - Some people live on less than $2 per day without clean water or access to education. How might their lives change under each of the above scenarios?

International Medical Corps, Pakistan

DFID/Vicki Francis

2. **Like people, countries demonstrate their value systems by how they spend money.**

   Read the words of the Director of the UN Millennium Campaign about money spent. (p.63) What does this disparity suggest about current societal values? Make a financial case for making different choices. Make a social justice case.

📖 Study the military budget for the US. 👁 What trends do you see? How does military spending compare with other US expenditures? Compare it with the military spending of other countries. What are the implications? Has it made the US any safer? How do you feel about what you have learned?

Mindmeister

3. Many scientists link population growth to sustainability concerns. Reactions to efforts to control population vary.

💡 Consider past efforts to control population and why they have succeeded or failed. To what extent would campaigns need to be modified for different societies?

➡ View a 5-minute video about population growth 👁 and this video by NPR. 👁 Sum up what you learned in the videos in a few sentences and share it with your community via social media or other means.

  • Check out the Population Connection website: *populationconnection.org*. 👁 How does watching this change your view on the problems and opportunities?

❓ Discuss the relationship between exponential population growth, resource distribution, and social justice. How has population growth put pressure on other areas?

4. There are many reasons why people migrate. (p.63)

➡ Volunteer for several hours over the course of a month at a center that supports immigrants.

➡ View a brief lecture by Robert McLeman on environmental migration. 👁

❓ Discuss the different kinds of migration and how they might change in the coming years as a result of environmental change.

➡ Attend a swearing-in ceremony for a new immigrant citizen. Interview someone about their recent immigration and citizenship journey.

➡ Share what you have learned with others in your community.

📖 Find out how many generations back your family came to the US and learn as much as you can about their experience. Ask who, what, where, when, and why.

The extra qualities needed now include intuition, empathy, feeling for nature, and sensitivity to spirit— in order for us to be able to map the new landscape appearing in front of us as the old way of being burns off like morning mist.

— *Hardin Tibbs*

Big Stock Photo

5. The role of women in curbing population growth is crucial.

➡ Watch the short video *The Girl Effect*. 👁

💡 Read the list of ideas about female empowerment. (p.65) Brainstorm other ways that world cultures could empower women to help encourage population stabilization.

💡 How would you educate adolescents to make wise child-bearing choices? What content or teaching strategies would you change depending on gender and culture?

6. The UN Millennium Development Goals offer an inspiring world vision (p.66-67)

♥ Divide your group in half. Each sub-group selects two of the goals and develops measurable indicators to track progress. Have a discussion afterwards.

➡ Collaboratively create a specific, measurable version of the UN Millennium Development Goals that meets the needs of your community. Discuss your rationale for selecting these goals.

7. What organizations are currently working to achieve a just society?

📖 Identify at least five organizations and select one you feel drawn to.

❓ Make a case for giving that group financial support.

➡ Use email or social media to help spread the word about this organization.

➡ Organize a fundraiser to bring awareness and resources to this cause.

8. How does privilege affect the way you live your daily life?

• Identify instances where privilege, lack of privilege, or blatant discrimination has affected the life of someone you know. These can be major life-shaping events like acceptance to a school, securing a job or business loan, or encounters with police. Identify other subtler situations—such as hanging around a park, service at a restaurant, or waiting in a line—where race or class seems to have played a role.

➡ Gather a diverse group of participants to take part in the Horatio

The irony of American history is the tendency of good white Americans to presume racial innocence. Ignorance of how we are shaped racially is the first sign of privilege.

— *Tim Wise*

Alger Exercise. 👁

➡ Take part in a school Challenge Day experience near you: sign up at *challengeday.org*. 👁

➡ Tell three other interested people about your experience and invite them to participate at the next opportunity.

9. Explore the distinctions between the following terms: indigenous people, aboriginals, native people, and first peoples.

📖 Read the thorough, fascinating treatment of these terms in the book *1491: New Revelations of the Americas Before Columbus*. Summarize what you learn at your next group meeting.

➡ Create a collage that captures your vision of equality, diversity, and intercultural harmony. Use photos from magazines that inspire you. Share this "dreamboard" with your friends in person or through photos.

10. According to Joanna Macy "The heart that breaks open can contain the whole universe."

♥ Exercise: Expanding Empathy

- Allow yourself to feel the suffering inherent in forced immigration, violence against women, and the plight of indigenous people.

- Connect with your empathy.

- Drop beneath your defense mechanism and allow your heart to go out to the mothers, daughters, sons, and fathers that the statistics represent.

- Identify one group of people in a specific place and picture their daily life.

- Imagine the fear, hope, despair, satisfaction, and creativity they may experience.

- Feel the strength and resilience they must have.

- Experience gratitude for their courage and connect energetically if you can.

📖 Search the Internet to find at least three examples of human interest stories that break down a stereotype or unexamined assumption. For example, consider the story of how an impoverished Mexican family sent generous donations to Hurricane Katrina victims in New Orleans.

> We lose our freedoms at the altar of fear.
>
> —*Kahlil Gibran*

Big Stock Photo

➡ View these human interest stories and the way in which they break down stereotypes: from CBS 👁 and CNN. 👁

❓ Make a point to cross lines of color, class, sexual orientation, and gender during the next week. Talk to at least one person who is of a different race or ethnic group. Ask them how their race/ethnicity has impacted their experiences with school, their workplace, or their community. Also, talk to at least one person from a significantly different social class, orientation, or gender and endeavor to gain a deeper understanding of how their life experience differs from yours.

❓ Discuss how discrimination based on color or race is different from or similar to discrimination that occurs due to class, gender, sexual orientation, or religious affiliation.

❓ What restrictions are faced by those imprisoned or living under an oppressive regime? Also, consider how a lack of literacy might affect the feeling of freedom.

> If we went back to the imprisonment rate we had in the early '70s, four out of five people employed in the prison industry would lose their jobs. That's what we're up against.
>
> —*Eugene Jarecki*

11. Review the Seven Foundations of a Just, Sustainable World (p.76-77)

➡ Choose one personal baby step that you will take in line with these goals. For example, if the goal you choose is "comprehensive peace," you could join an event put on by Code Pink, or you could commit to meditate five times a week for a month.

❓ What role could you play in revitalizing your community? Discuss. Find out what organizations there are in your community that are making a difference and see how you can participate or support them.

❓ Do you feel you have economic and social privilege, based on your race or other factors that you inherited? What ways to you feel you could contribute to a more equitable society?

12. The US has a higher rate of incarceration than nearly every other country.

📖 How are recidivism rates different when services are offered versus when they are not? Look at the impact of 1) addiction treatment programs provided to prisoners and/or 2) community support programs provided after release.

➡ "Success for Life" exemplifies how such programs are implemented and the benefits they provide; see the video. 👁

The American Psychological Association published findings about the efficacy of drug treatment programs in prisons. ◉ How could this information be implemented into the current justice system?

The Center for American Progress provides commentary on the ways in which race factors in to the justice system. ◉ How might such a flawed system be remedied?

➡ Visit a prison and speak with inmates about their experiences "behind bars," or speak with a former prisoner.

➡ If there is a prison nearby to where you live, find out if there are service programs being offered in which volunteers go in and offer services to the inmates—for example, discussion groups, meditation or yoga classes, etc.—and see if you can participate.

➡ Correspond with a prisoner and ask questions about his or her experience and expectations for a future "life on the outside"—go to *prisonpenpals.com* (note that you must be 18 years or older to engage in a prison penpal correspondence.)

Part Five: ECONOMICS
What Counts: Valuing Life

Dazzia Szczepaniak

Confounded by the illusion that our world is one of endless open frontiers with abundant resources free for the taking, we humans have created an economic system shaped by rules that… reward individualistic competition, material accumulation, and reckless consumption.

We are just beginning to come to terms with the reality that we inhabit a living spaceship of finite resources and intricately balanced ecosystems… Managing economies to maximize growth … on a finite planet is the equivalent to maximizing the rate of consumption of essential resources on a spaceship.

Both are actively suicidal.

— *New Rules for a Spaceship Earth,*
New Economy Working Group

In our "growth society," success—defined by ever-increasing profits—is measured by Gross Domestic Product. GDP values the disastrous Deepwater Horizon oil spill infinitely more than it values caring for your child or grandparent. Oil spills cause money to circulate in the economy. Caregiving for family does not.

Shifting from this "growth society" to a life-sustaining human presence on the planet is our current imperative. It's also an opportunity to reinvent our cultures, our institutions, and ourselves.

We can actualize new rules for "… a regenerative society [that] is a flourishing society. The revolution is not about giving up; it's about rediscovering what we most value … making quality of living central in our communities, businesses, schools, and societies. It is about reconnecting with ourselves, one another, and our fellow non-human inhabitants on Earth."[1]

Here are a few suggestions for these new rules:

Surf the energy flux. Live within our energy income by relying on renewable sources of energy, such as solar, wind, and environmentally sustainable hydropower.

Zero to the landfill. Everything, from cars and iPods to office buildings and machine tools, should be treated as 100% recyclable, re-manufacturable, or compostable. Nature produces zero waste. We forget that we nature.

We are borrowing the future from coming generations; we must pay it back. Our first responsibility is to leave a healthy global biosphere for our children, their children, and on into the future.

We are only one of nature's wonders. Just one of countless species, all of which matter. We depend on each other in ways we cannot even imagine.

Value the Earth's services. Healthy ecosystems are the source of all life. What we do to them, we do to ourselves.

Embrace variety and build community. Harmony amid diversity is a feature of healthy ecosystems and societies.

In the global village, there is only one boat, and a hole sinks us all. Our mutual security and well-being depend on respect and concern for all. If any of us is insecure, then we all are.[2]

"A crisis is an opportunity riding the dangerous wind."
—*Chinese proverb*

GLOBAL CRISIS = OPPORTUNITY FOR CHANGE

In *The Great Turning: From Empire to Earth Community*, David Korten refers to our global crises as humanity's "defining moment of choice between moving ahead on a path to collective self-destruction and joining together in a cooperative effort to navigate a dramatic turn to a new human era."[3]

The New Economy Working Group *(neweconomyworkinggroup.org)*, co-chaired by Korten and John Cavanagh, director of the Institute for Policy Studies, affirms that the recent financial crisis "has put to rest the myths that our economic institutions are sound. Financial failure is just the tip of the iceberg. Their grossly inefficient and unjust allocation of the resources necessary for human well-being bears major responsibility for spreading social and environmental collapse."[4]

Korten's analysis looks squarely at economic and social justice linkages—an essential aspect of the emerging green economy that has been a long time coming: "Equalizing economic power and rooting it locally shifts power to people and community from distant financial markets, global corporations, and national governments. It serves to shift rewards from economic predators to economic producers, strengthens community, encourages individual responsibility, and allows for greater expression of individual choice and creativity."[5]

According to Korten, "Real wealth is created by investing in the human capital of productive people, the social capital of caring relationships, and the natural capital of healthy ecosystems…. Markets have a vital role, but democratically accountable governments must ensure that everyone plays by basic rules that internalize costs, maintain equity, and favor human-scale local businesses that serve community needs."[6]

To make the most of current opportunities and changes, we must summon the collective will to, in Korten's words, "create a money system that serves people, community, and the whole of life. To do so we liberate our minds from the illusion of a cultural story that the making of money is synonymous with creating wealth."[7] In the long run, a people-based economic system that equitably supports human health and well-being will be a much more sustainable system.

Bring On the New Economy!

To reform economic policy toward the mission of sustainability, the New Economy Working Group, whose focus is on bringing social and environmental issues to economic policy discussions, implores that we support localization initiatives to rebuild "Main Street." This neighborly term refers broadly to economies made up of a collection of human-scale enterprises devoted to serving the needs of people and nature—enterprises that are fully accountable to the community.

We need to reduce aggregate consumption and reform the way we measure economic performance, using health, equity, and well-being as indicators. We need to support political action to convert the inequitable Wall Street money system and war economy to one that is equitable, supports peace, and provides direct benefit at the community level.

To bring Earth's life support system back into balance, we must reallocate our use of finite resources to beneficial and more efficient uses. A just redistribution of wealth—both income and ownership—will help secure the health and general well-being of all. We won't be able to simply buy ourselves out of economic crises and environmental collapse. More than stimulus and recovery programs, we need to restructure local, regional, and global rules and institutions. We can start with our own communities.

Globalization

The wealthiest one percent of the population now owns 46% of the world's wealth.[8] Increasing concentration of financial power and economic growth hastens the destruction of the entire life-support system of the Earth. The evidence of systemic failure includes unemployment, displacement of indigenous peoples, and massive immigration.

Most social and ecological costs, such as pollution and natural disasters, are still externalized—in other words, they are not included in the prices of goods and services and are not paid for by industry, but indirectly by the public and the biosystem.

Brief History: Background on Globalization and Free Trade. In the book *Alternatives to Economic Globalization*, the International Forum on Globalization explains "since World War II, the driving forces behind economic globalization have been several hundred global corporations and banks that have increasingly woven webs of production, consumption, finance, and culture across borders... These corporations have been aided by global bureaucracies that have emerged over the last half-century, **with the result being** a concentration of economic and political power that is increasingly unaccountable to governments, people or the planet and that undermines democracy, equity, and environmental sustainability."

The current international banner-carrier of deregulation, the World Trade Organization (WTO), was set up to prevent obstacles to global commercial interests. Many polices used to achieve these goals, however, effectively undermine working people, labor rights, environmental protection, human rights, consumer rights, social justice, local culture, and national sovereignty.

In order to assist the rebuilding of Europe after World War II, finance ministers and heads of corporations and banks convened in 1944 in a historic session in Bretton Woods, New Hampshire. They set up the centralized World Bank and International Monetary Fund (IMF). Devised to accelerate worldwide economic development and stabilize currency exchange rates, the IMF focused on bringing "underdeveloped" countries into the global economy.

In 1948, the General Agreement on Tariffs and Trade (GATT) began regulating manufactured-product trade quotas — later adding investment and corporate services trading. GATT folded into the WTO in 1995. Critics of the WTO point out that world trade rules continue to be set behind closed doors by corporate interests, circumventing democracy and excluding the most harshly impacted countries as well as all non-governmental organizations (NGOs).

The Center for Food Safety reports that the WTO and other free trade systems like the North American Free Trade Agreement (NAFTA), have "eliminated a nation's rights to protect its citizens and its natural resources, while allowing multinational corporations uncontrolled and unrestricted access to a country's markets and resources.... California's farmers and rural communities are disappearing; crops that once thrived and were profitable are now being plowed under; and ecosystems are collapsing at alarming rates."

According to The Center for International Policy, small farmers in Mexico "are being driven off the land, forcing an increased reliance on imported resources. Meanwhile, major US agribusiness firms have grown by leaps and bounds under the auspices of the free trade model."[9]

Trade agreements that deregulate corporations

and financial markets, along with banking consolidation, are now widely seen as having paved the way for the recent financial crash. Affecting far more than trade, these instruments are about "eliminating the regulation of corporations, eliminating public services, and putting public assets up for sale to the highest bidder while holding governments accountable for enforcing intellectual property rights monopolies," David Korten says.

The "Old Economy" also privatizes the commons and reduces or eliminates taxes on the rich. This privatization of the commons has prevented local communities from benefitting from the resources on their own lands.

Global Solutions

Solutions lie in systemic change driven by individual actions, localized systems, organized communities of place and interest, and collaboration on a global scale, while dramatically restructuring global rules and institutions. The groundswell of new ballot initiatives and the expansion of long-standing models are awe-inspiring, springing from a wide diversity of people, communities and institutions.

For example, in the US, the Citizens Trade Campaign *(citizentrade.org)*, promotes legislation to "address the needs of ordinary Americans on issues such as jobs, wages, the environment, agriculture, consumer safety and public health" and create a trade agenda that "advances democracy and economic development from the bottom up." Its goal is to reform policies to "reward the work that creates wealth, with meaningful protections for our people and our planet."[10]

Hazel Henderson, economist and founder of Ethical Markets Media, advocates creating an economy aligned with true democratic access, resulting in a better world: "As world trade evolves into exchanging what works, as well as continuing to savor and trade cuisine, art, music, dance and literature, world trade can actually help our human family evolve toward higher levels of planetary awareness and Earth ethics."[11]

Numerous groups and economists are calling for action to support this type of change. You can too. Tell lawmakers that you support actions such as these:

- Follow principles of UN agreements that support human rights, labor standards, indigenous rights,

Vietnam microfinance project, supporting small farmers and vegetable sellers

The Movement to End Corporate Personhood

"Corporate personhood" is the controversial legal designation in which corporations are given the same rights as individual persons under the law. Many economics and social justice activists believe that ending corporate personhood is key to creating a just society and a healthy economy.

The seed for corporate personhood was planted in 1866 when the Supreme Court decided the case of *Santa Clara Co. v. Southern Pacific Railroad.* The Court ruled that, since corporations are groups of people acting as one, a railroad company was a "person" in the eyes of the law, and thus subject to the protection afforded by the Fourteenth Amendment to the US Constitution. This amendment, written after the emancipation of slaves, forbids any state from denying any "person" equal protection under the law.

However, assignment of corporate personhood carries some quite controversial protections:

- Protection against routine inspections, unless the inspector has a warrant or prior permission.

- Protection against full disclosure of what their products are made with.

- The ability to use unlimited corporate money on ballot initiatives and referenda which could effectively work against the common good (since the common good is often at odds with the short-term interests of corporations and their shareholders).

- Protections for chain stores that were intended for small, local businesses.

Together, these protections make it hard for citizens to hold their own against a corporation. Small businesses and individual citizens are impossibly out-gunned on issues related to taxes, legislation, political campaigns, the environment, social justice, small-business survival, and other impacts of corporations upon their communities.

In recent years, a citizens' movement has arisen to amend the constitution to end "corporate personhood." This movement has been growing in popularity since the 2010 Supreme Court ruling in the case of *Citizens United v. Federal Elections Commission.* The Court ruled that governments may not restrict political expenditures by corporations, labor unions or associations.

Many groups have been working toward this goal of a Constitutional end to corporate personhood -- among them, Move to Amend (*movetoamend.org*), Free Speech for People (*freespeechforpeople.org*), Alliance for Democracy (*thealliancefordemocracy.org*), The Community Environmental Legal Defense Fund*celdf.org*), The Women's International League for Peace and Freedom (*wilpfus.org*), The American Independent Business Alliance (*amiba.net*), and The Program on Corporations, Law and Democracy (*poclad.org*).

Action within these groups generally starts at the local level. The idea is to build a strong enough grassroots movement that Congress will be forced to notice and act upon its demands. Many states now have campaigns in progress. As of 2013, according to Move to Amend, nearly 20 states had sent Constitutional amendment proposals to Congress; the momentum is surely building now.

— Beth Vanden Heuvel,
Tri Marine International

and environmental protection.

- Reform international finance institutions and rules to support social equity and environmental responsibility.

- Reform the global governance framework to support human rights, including labor rights and environmental protection.

- Institute responsible lending and expand debt cancellation for developing countries.

- Reform corporate subsidy systems and institute corporate accountability.

- Redefine indicators of economic progress to include environmental and social impacts, and insist on transparency.

- Eliminate the need for natural resource extraction linked to conflict, war, and civil strife.[12]

Fair Trade Certification

A market certification strategy, Fair Trade (*fairtrade.net)* ensures that trade relationships benefit local communities, workers, and the environment from which commodities originate. A price premium is set to cover costs *plus* investment in education, healthcare, and ecosystem restoration.

Worker empowerment and collectives of small producers are essential components of the Fair Trade industry. NGOs promoting Fair Trade create online marketplaces, facilitate associations (World Fair Trade Organization and Fair Trade Federation), and provide education (Fair Trade Resource Network). To ensure that standards are met before products receive the Fair Trade stamp, a rigorous evaluation is performed by one of the 25 members of Fairtrade Labeling Organization International (FLO).

When you buy crafts, clothes, cotton, coffee, tea, sugar, chocolate, honey, wine, fruits, spices, grains, or flowers — all products grown under a Fair Trade certified label in many countries in Africa, Asia, or Latin America — you are making a difference in alleviating poverty and protecting ecosystems.

Fair Trade can work wonders. A banana plantation in Ecuador, a tea producer in Kenya and a cot-

Organic, fair trade coffee cherries waiting to be processed. San Pablo Farm, Chiapas, Mexico

ton farmer in Mali can now send their children to school and access health clinics. Fair Trade has also enabled farmers to rebuild bridges after a hurricane and develop water, food storage, and sanitation systems. One African farmer said, "If the whole value chain was made fairer, Africa would be lifted out of poverty."[13]

Forest products' certification similarly helps protect forests from destruction. The Forest Stewardship Council (FSC) is a leader of this market-based approach. There are also multiple certification systems for wild harvesting of medicinal and aromatic plants, including, FSC, Fair Wild, and Organic Wild Crop.

Credit to the People, Global to Local

Can a small investment make a big difference? According to advocates of microcredit, the answer is a resounding "yes."

Microcredit is the making of small loans to impoverished people—frequently women—who lack access to more traditional financial services. The loans make it possible for them to start entrepreneurial enterprises.

Grameen Microcredit — Inspiring Microfinance Institutions Around the World: Often cited as the first microcredit organization, Bangladesh's Grameen Bank (GB) starts with the belief that credit is a human right. They lend money to people based on their potential, regardless of their level of material possessions.

More than 90% of the borrowers are women. Repayment rates average 97%. GB is owned by the borrowers and provides customized credit in diverse countries, economies, and cultures. One study reports that 68% of borrowers' families have crossed the line out of poverty. Higher education loans and scholarships help the next generation of Grameen families.

US President Barack Obama said of the 2006 Nobel Peace Prize recipient and Grameen founder, Muhammed Yunus, "He revolutionized banking to allow low-income borrowers access to credit…. He has enabled citizens of the world's poorest countries to create profitable businesses, support their families, and help build sustainable communities. In doing so, he has unleashed new avenues of creativity and inspired millions worldwide to imagine their own potential… [The Nobel Committee recognized microcredit as] an important liberating force in societies where women in particular have to struggle against repressive social and economic conditions."[14]

Peer to Peer: Launched in 2005 to help alleviate poverty, Kiva *(kiva.org)* provides a person-to-person micro-lending website, empowering individuals to lend directly to unique entrepreneurs around the world. Lenders can invest any amount between $25 and $10,000 to projects they choose online. While there's no profit to investors, repayment rates are nearly 100%. Kiva's field partners, microfinance institutions that provide localized program support, manage the loans.

"This is a moment in history when the average person has more power than at any time."

—Katherine Fulton, president, Monitor Institute
(community philanthropy)

ENTERPRISE ECONOMICS

All over the world, millions of people are redefining economic relationships within their communities. This work is vital to the creation of a counterforce to the old economy, providing at least a measure of balance that supports local, small-scale and individual pursuits.

Renowned environmentalist David Suzuki says that we must no longer "maintain a conceit that we can manage our way out of the mess, increasingly with heroic interventions of technology…. We've learned from past technologies—nuclear power, DDT, CFCs (chlorofluorocarbons)—that we don't know enough about how the world works to anticipate and minimize unexpected consequences….We have no choice but to address the challenge of bringing our cities, energy needs, agriculture, fishing fleets, mines, and so on into balance with the factors that support all life. This crisis can become an opportunity if we seize it and get on with finding solutions."[15]

The New Economy Working Group asserts that we need to "favor organization forms that lend themselves first to serving the community and treat financial return as a secondary consideration. The preferred models are small to medium-sized community-rooted cooperatives, worker-owned, community-owned, and various locally-owned independent businesses." These structures can also "aggregate economic resources" through alliances of many forms, such that they "do not create concentrations of monopoly power or encourage absentee ownership."[16]

Nature's Design

Halting the push for relentless economic growth is the most effective way to conserve and maintain biodiversity because "as the economy grows, natural capital, such as air, soil, water, timber, and marine fisheries, is reallocated to human use via the marketplace, where economic efficiency rules." Nature's lessons suggest that we must "favor bio-physical effectiveness over economic efficiency," and repair the ecosystems upon which life depends.[17]

The natural world models what works. Innovators in the field of biomimicry are looking closely at nature's designs and using its principles. Author and biomimicry pioneer Janine Benyus says: "Just as we are beginning to recognize all there is to learn from the natural world, our models are starting to blink out — not just a few scattered organisms, but entire ecosystems. A survey by the National Biological

Service found that one half of all native ecosystems in the United States are degraded to the point of endangerment. That makes biomimicry more than just a new way of viewing and valuing nature. It's also a necessary race to the rescue."[18]

Biologists with the Biomimicry Guild *(biomicry.net)* help companies, cities, nonprofits, and agencies create products, processes, and policies using nature's time-tested patterns and strategies. Products resulting from the application of biomimicry include a solar cell inspired by a leaf, self-cleaning surfaces, non-toxic dyes and adhesives, passive home cooling, and pumps that replicate the motion of seaweed.

Full Accounting

No-Waste Accounting: In our current economic model, industry does not have to account for impacts on the planet or its living communities—impacts from extracting resources to make products, or from the waste and pollution produced in the production, distribution, use, and eventual disposal of the product being produced. These hidden, "externalized" costs don't show up in the accountant's "costs" of producing the product, or therefore in its price.

Enterprises seeking integrity realize all these costs must be "internalized." Using a "cradel to grave" linear model, they analyze all the real costs in the steps along their supply chain. For a summary of the process, watch the Story of Stuff video *(storyofstuff.org)* that illustrates how the supply chain affects everything.

Improving eco-effectiveness requires full-cycle accounting—epitomized by Cradle to Cradle *(c2ccertified.org)* design—in which industrial processes don't generate waste or toxic pollution, just like nature.

With this model, "cyclical material flows...like the Earth's nutrient cycles, eliminating the concept of waste." Each part of a product is designed to be safe and effective, and to provide high-quality resources for subsequent generations of products. All material inputs are conceived as nutrients, circulating safely and productively, so products are either completely recyclable in the "technosphere" or become biodegradable food for the biosphere.

Market-Based Approaches

While our current economic system is one of the major drivers of environmental degradation, the power of that same system can also be harnessed for good. How? By regulating taxes, prices, and incentives in ways that encourage people and companies to act in environmentally and socially beneficial ways.

What's the problem? It balances, doesn't it?

One example of this is the emerging field of ecosystem services. This term refers to the placement of economic value on the benefits humans derive from the Earth's natural processes, such as a wetland's water purification process, or trees sequestering carbon.

Payments for ecosystem services could be monetary incentives or subsidies provided to individuals, organizations or communities that use the resources on their property and land responsibly.

One example of a successful Payment for Ecosystem Services (PES) project was one that protected the New York City watershed. The city wanted to protect its drinking water through conservation of its watershed. Instead of building a costly water filtration plant, the city opted to buy thousands of upstate acres, shield its reservoirs from pollution, improve treatment plants and septic systems, and subsidize environmentally sound economic development.

The main beneficiaries were the farmers who relied on the Catskills watershed. They received technical and managerial assistance, new farming equipment, and infrastructure improvements to their agricultural operations. The program also paid them to remove sensitive streamside lands from agricultural production, and rewards them for their long-term commitment to sustainable agriculture. Other successful PES programs have been implemented around the world; they are becoming more viable as developing countries realize the importance of sustaining precious ecosystem services.

Corporations Going Green: Incentives and Progress Measures

Progress toward a greener economy is being made by consumers and businesses focused on sustainability, by socially responsible investors, and via corporate social responsibility, but many contend that the shift is too slow. Many doubt whether corporations publicly traded on a stock exchange—corporations legally bound to focus on profits for shareholders—can ever measure up to even the most basic definition of sustainable development: "meeting the needs of the present without compromising the ability of future generations to meet their own needs."[19]

Given this mandate to maximize profits, publicly traded corporations tend not to invest in environmental and social considerations. Due to this constraint, Seventh Generation, a pioneering maker of

Green Festival, San Francisco; helping to promote the market for environmentally-friendly products

safe and environmentally friendly household products, reverted back to private ownership after seven years of being publicly traded. Loss of control was a major factor cited in the decision to leave the market. The shareholder focus on short-term profitability, maximizing the value of stock, and showing corporate growth in order to attract investors were all challenges to an ethic that did not center on profit.[20]

And at Clif Bar & Company, even though their largest competitors, Power Bar and Balance Bar, are now owned by Nestlé and Kraft, owner Gary Erickson decided not to sell, in order to ensure that his progressive vision stays alive and high organic standards are maintained.[21]

Many companies that disingenuously tout supposedly "green" measures and publish slick sustainability reports (see "Greenwashing" below) are "really focused on [maintaining] a flawed economic development system that is increasingly based upon the addiction to commodified material consumption," says John Ehrenfeld, executive director of the International Society for Industrial Ecology. He calls for a radical transformation in thinking and action that acknowledges a "deep-seated systems failure."[22]

To make it financially feasible for corporations to internalize costs, they need a more level playing field, through government rules and incentives or a different corporate structure. Otherwise, David Korten explains, they risk takeover or being "driven from the market by competitors that gain a market advantage by externalizing these costs, as, for example, Wal-Mart," which relies on cheap supplies and low wages.

To institute change, critics of corporate influence suggest strategies such as government regulation, grassroots action, exposing socially and environmentally damaging corporate practices, and building viable alternatives to the corporate-dominated economic system.

Corporate Social Responsibility (CSR)

While many question the extent to which reform is possible within the current system, some activists, consultants, and corporate leaders are making progress in promoting and instituting Corporate Social Responsibility (CSR).

The blossoming "conscious capitalism" movement,[23] featuring high-profile companies like Whole Foods, The Body Shop, and Patagonia, demonstrates the potential of profitable businesses "doing well by doing good." They reduce resource and energy consumption, attract sustainability- and social-responsibility minded MBA graduates and investors, satisfy green consumer demand, and avoid negative publicity.

Some companies are adding hybrid and biodiesel vehicles to their operations, reducing toxics used in manufacturing, conserving water, buying recycled plastics, and conserving energy. FedEx is finding more fuel-efficient airplanes and using smaller vehicles.

Companies "doing good" often explicitly embrace the Triple Bottom Line (TBL)—people, planet, and profit. Evaluating progress since he coined the term in 1997, corporate responsibility expert John Elkington concluded that "the TBL agenda as most people would currently understand it is only the beginning." He recommends policies on taxation, technology, labor, and corporate disclosure involving diverse stakeholders.[24]

Some highly ethical companies are now becoming "B Corporations" (bcorporation.net). This third-party certification designates companies that meet environmental and social responsibility standards. The B Corp designation is comparable to that of Fair Trade certification for products.

Governments can play a role beyond the regulatory realm by demanding that corporate goods, services, operations, and overall performance meet criteria set up through credible indicator systems, including requiring companies to take full responsibility for environmental and social costs.[25]

Of the hundreds of initiatives that offer parameters for organizations seeking to embody sustainability, three prominent frameworks — the Earth Charter, the UN Global Compact, and the Global Reporting Initiative (GRI)—are voluntary, partnership-based frameworks derived from international norms that help quantitatively measure adherence to CSR.[26]

The Earth Charter is compatible with the UN Millennium Development Goals. CSR reports that follow the GRI disclose a company's economic, social, and environmental impacts based on indicators including water and energy use, waste and emissions, labor practices, product safety, human rights, and community impacts.

Investing in CSR: Ceres *(ceres.org)* is a large US-based network of investors and environmental and public interest groups that works with companies and investors to integrate sustainability into capital markets. Its recent analysis concluded that while many companies "are making progress, their actions to date are only the beginning of what is needed."[27] Among their recommendations is that companies should tie compensation packages to climate performance measures, which would help to increase engagement.

Socially Responsible Investing (SRI): A growing number of investors, pension funds, agencies, and collectives pay close attention to the environmental and social impact of where their money goes. A leading source of information about SRI for the US, *(greenamerica.org)* offers guidance, links to resources, and community investing opportunities. Green America's Green Business Network features about 5,000 carefully screened small to medium-sized business members. Green America's *National Green Pages* is an annually published list of these ethical, responsible businesses.

The Investors' Circle (IC) Foundation *(investorscircle.net)* has supported more than 270 companies with $170 million, investing in energy, health, education, and media, as well as community development and minority- and women-led enterprises. Investors' Circle has also pledged more than $4 billion in follow-up investment, which allows these companies to further develop.

The IC set up a new nonprofit in 2008 called the Slow Money Alliance *(slowmoney.org)*, which aims to re-localize money and jobs, thereby countering the corporate culture's tendency toward speed, quantity, and growth. Slow Money advocates promote the expansion of enterprises that combine nonprofit and for-profit goals.[28]

"The individual is forcing the change. People are shopping around, not only for the right job but for the right atmosphere. They now regard the old rules of business as dishonest, boring, and outdated. This new generation is saying, 'I want a society and a job that values me more than the gross national product. I want work that engages the heart as well as the mind and the body, that fosters friendship and that nourishes the Earth. I want to work for a company that contributes to the community.'"

—Anita Roddick, founder, The Body Shop

Greenwashing: Citizens and environmental groups paying attention to corporate marketing messages warn of pervasive greenwashing. The US-based research group CorpWatch defines greenwashing as "the phenomenon of socially and environmentally destructive corporations attempting to preserve and expand their markets or power by posing as friends of the environment."[29]

While there can be truth to some of the positive spin, the average consumer is probably unaware of what lies behind it. For example, Chevron has underwritten public TV programs that carry ads touting the oil giant's investments in green energy and innovation.

At the same time, "Over three decades of oil drilling in the Ecuadorian Amazon, Chevron dumped more than 18 billion gallons of toxic wastewater into the rainforest, leaving local people suffering a wave of cancers, miscarriages and birth defects,"[30] reports independent filmmaker Joe Berlinger in his award-winning documentary *Crude*.

Meanwhile, the international campaign seeking justice in Ecuador, ChevronToxico *(chevrontoxico.com)*, launched by Amazon Watch, is urging consumers, shareholders, and the public to take action, support the people and communities of the rainforest, and pressure Chevron to clean up.

Greenpeace *(greenpeace.org)* hosts an information, news, and action website about greenwashing that explains how consumers suspecting misleading advertising can contact the US Federal Trade Commission (FTC). In addition, guides for the use of environmental marketing claims have been revised by the FTC as they evaluate terms such as eco-friendly, sustainable, and carbon neutral.[31]

To help consumers evaluate marketing claims,

hold businesses accountable, and stimulate the market demand for sustainable business practices, EnviroMedia Social Marketing and the University of Oregon set up a Greenwashing Index online *(greenwashingindex.com)*.

Climate Washing: The most blatant examples of this new PR trick are seen among corporations claiming to support greenhouse gas emissions reductions, while at the same time paying lobbyists and trade associations to defeat such policies and legislation—an obvious double-standard.

Local Washing: Some food corporations have gained notoriety for "local washing" (for example, Hellmans, Unilever, Frito Lay, Foster Farms) by making misleading claims about product sourcing, while a giant bank, HSBC, inaptly calls itself "the world's local bank."

Even Walmart, Winn-Dixie, and Starbucks are getting in on the "local" act.[32] As a result, some true localization efforts are becoming more specific with their claims: "locally-owned and *independent* businesses." Vermont has strict rules about this type of claim and other places could follow their lead using the model that Vermont has pioneered.

Researchers with the Institute for Local Self-Reliance *(ilsr.org)* report that communities with buy-local campaigns fare better financially, and that "in city after city, independent businesses are organizing and creating the beginnings of what could become a powerful counterweight to the big-business lobbies that have long dominated public policy. Local business alliances have formed in more than 130 cities or states and together count some 30,000 businesses as members."[33]

To advance localization campaigns, communities can get support from the American Independent Business Alliance and the Business Alliance for Local Living Economies (BALLE; *bealocalist.org)*.

LOCAL SOLUTIONS
The Earth Community Economy

"The communities with the best prospect to weather the mounting forces of a perfect economic storm will be those that act now to rebuild local supply chains, reverse

Demonstration in Richmond, CA, site of a major Chevron oil refinery

the trend toward conversion of farm and forest lands, concentrate population in compact communities that bring home, work, and recreation in easy reach by foot, bicycle, and public transportation, support local, low input family farms, and seek to become substantially self-reliant in food and energy."

—*David Korten*, The World We Want

Cooperation is key to building the local economy, which requires restoring caring relationships and finding common values.

Sustainability advocate Chris Maser writes: "To protect the sustainability of a resident's community within a landscape, the community's requirements must be met before other considerations are taken into account; if this does not happen, no other endeavor will be sustainable.... The choice of how and why we alter the Earth is ours, the adults of today. The consequences we bequeath to every child of today and beyond. How shall we choose—to protect the commons as the unconditional gift of Nature that is everyone's birthright or continue to fight over how we are going to carve it up for personal gain and so despoil it unto everlasting?"[34]

Bioregional solutions include learning from and applying healthy traditions (ecological knowledge, permaculture, organic farming, appropriate technology, low-impact design, seed swaps, and other "back to the land" strategies), reclaiming community support, and rebuilding the resiliency we need to adapt to change. Prevalent among rural strategies we find a restoration-based economy in natural resource sectors, such as sustainable forestry, fishing, and agriculture — including enterprises involved in urban-to-rural market linkages.

Smart Revolution

Localization: Creating and shoring up local economic systems and resources reduces local economic vulnerability, increasing social and economic resilience in uncertain times. A focus on the local can minimize the negative social and environmental externalities that come with inefficient trade and absentee ownership that supports and is supported by Wall Street-type financial institutions.

Spending money in your neighborhood creates the local multiplier effect—measured in jobs, income, tax revenues, and wealth. Local businesses are essential to healthy tourism, participation in politics, charities, and social activities.

Michael Shuman (author of *The Small-Mart Revolution: How Local Businesses Are Beating the Global Competition* and *Going Local: Creating Self- Reliant Communities in a Global Age*) explains that spending locally helps communities plug "leakage" of money out of the local system.

Dollars leave the community unnecessarily when imported products and services could be sourced locally. Other than new electronic devices, appliances, cars, and fossil fuel, there are few things that even a rural community could not supply its residents — provided the local independent businesses exist and are supported. The more money that circulates in an area, the more income, wealth, and jobs it generates. Compared to spending money in a chain or franchise business, a dollar spent at a local restaurant has 25% more impact—60% for a retail shop, and 90% for services.[35]

The Small-Mart Revolution Checklist *(small-mart.org/files/SMChecklist-Globalizers.pdf)* offers 96 such suggestions. Ones that can save money include "honor junk," use local currency and barter systems, and encourage renting over buying. Its "daily mantra" for policymakers: "Remove all public support, including anything that requires city staff time and energy, from nonlocal business and refocus it instead, laser-like, on local business."

Localization enhances equity and stewardship as well, as it encourages people to come together to care for their communities, neighbors and their environment.

Local Empowerment

You can help ensure public interest is served, citizens are valued, and basic human rights are met. Among "31 Ways to Jump Start the Local Economy," *YES!* Magazine suggests that you:

- Help folks cope in economic downturns and act together to create the new economy: start a Common Security Club in your faith community or neighborhood.

- Reach out to groups that are organizing people on the frontlines of economic crises, like Jobs with Justice and Right to the City.

Beddington Zero Energy Development is an environmentally friendly housing development in Hackbridge, London

- Keep your energy dollars circulating locally. Launch a clean energy cooperative to install wind turbines or solar roofs, and to weatherize homes and businesses.

- Declare an end to corporate personhood in your community.

Communities demanding to decide for themselves what industry and outside corporations can or cannot do have passed ordinances that ban destructive corporate practices. Given the reticence of the US Congress to do so, local efforts are critical.

Community-Friendly Enterprises

The Small-Mart blog offers information on legal structures for stimulating investment in local businesses and entrepreneurs, including micro-lending and cooperatives.[36]

Besides finance, food, and energy sectors, local economic development can focus on recycled and reclaimed products, entertainment, and healthy lifestyles and services.

Social enterprises focus on public benefit and community service. The financial return is a means for meeting community needs rather than private profit. The Social Enterprise Alliance for Midlothian in Scotland includes business-like enterprise, cooperatives, and credit unions, as well as traditional non-profits, in their definition of social enterprise.

Associations of small businesses and entrepreneurs are more resilient than individual competitors. For example, farms can cooperate in community-supported agriculture (CSA) programs offering a wider variety of products to their customers under one umbrella—adding value to the benefits of individual farm CSAs. Limited liability partnerships are a newer model than consumer, producer and worker cooperatives, are fairly flexible, and can combine aspects of cooperatives with other goals.

"Good society" advantages of innovative locally-owned enterprises, according to the National Center for Economic and Security Alternatives, include their emphasis on democratic values, accountability, and local control of assets, and capital that remains in the community. This group tracks many emerging forms of community-rooted, asset-building institutions and enterprises such as community development financial institutions (CDFIs), municipal and nonprofit enterprises, urban land trusts, and local currency and barter systems.

Community-Wealth.Org hosts a directory of community-building resources, covering many types of financial institutions and municipal initiatives, socially responsible investing, partnerships, transportation-oriented development, and more. There's no shortage of creative ideas and opportunities, particularly when groups join forces to use tried-and-true principles with the new wave of expanded technology applications.

COOPERATIVES
Membership-Based Banking and Business

One member, one vote.
                    —A co-op principle

Cooperatives subscribe to seven principles:
1. voluntary and open membership
2. democratic member control
3. member economic participation
4. autonomy and independence
5. education, training and information
6. cooperation among cooperatives, and
7. concern for community

The three main types of cooperatives are consumer cooperatives (including credit unions and food coops), producer cooperatives (such as agricultural producers), and employee-owned cooperatives.

Credit unions—local, member-owned and democratically governed—invest in their members and communities while offering standard banking services.

The World Council of Credit Unions *(woccu. org)*, with 177 million individual members served by 49,000 credit unions in 96 countries, is among a growing number of institutions that have revolutionized micro-enterprise finance to improve incomes in developing countries.

To raise awareness about the challenges that women face daily in accessing financial and other resources, the council's Global Women's Leadership Initiative works in partnership with the Canadian Co-operative Association that was started in 1909.

The International Cooperative Alliance *(ica. coop)*, with 223 member organizations from 87 countries, representing 800 million people who participate in the cooperative movement, reports that locally controlled, fiscally conservative credit unions and co-operative banks are faring well. The rate of new co-op formation of all varieties is rising.[37]

In rural Mississippi, Winston County Self Help Cooperative brings small farmers and landowners together to overcome adversities. When they started in 1985, "small family farmers were under siege due to unfavorable financial conditions and USDA's lack of interest in serving black farmers, [who] needed an outlet to earn more income from products..."

The Self Help Cooperative adopted a mantra of "Saving Rural America" to emphasize the stewardship of their natural resources and support long-range planning and farm diversification. Key partners include the Federation of Southern Cooperatives and Heifer International.

The Mondragon Corporation is a corporation and federation of worker cooperatives based in the Basque region of Spain

INNOVATIVE COMMUNITY DEVELOPMENT BANKS

"Communities cannot achieve economic prosperity if entrepreneurial activities and residents' health are compromised by toxins in the land, air and water, or if natural resources are consumed in an unsustainable way."

—*ShoreBank (now part of Urban Partnership Bank)*

New Resource Bank is a San Francisco, California-based bank with a mission to advance sustainability on every level: the loans it makes, its own operations, and its engagement with the community. It offers a range of commercial loan and deposit products, with expertise in renewable and alternative energy, green building, organic food, and green products and services.

The innovative bank was founded in 2006 with a vision of bringing new resources to sustainable businesses and ultimately creating communities that are more self-sufficient and resilient. All new loan recipients must be green businesses or committed to improving their operational sustainability and managing their impact on society and the environment.

In 2010, New Resource became a B Corporation. B Corporations meet comprehensive social and environmental performance standards; the bank was the first publicly traded company to sign on.

Lakota Funds

Modeled after the Grameen Bank microcredit structure (see "Global Solutions" above), Lakota Funds *(lakotafunds.org)* serves the Oglala Lakota Oyate (People) on the Pine Ridge Indian Reservation in South Dakota with business loans, technical assistance, leveraged funding from outside sources, and facilitation for groups of borrowers. This first Native American Community Development Financial Institution helps to break the cycle of poverty.

When Lakota Funds started there were only two Native American-owned businesses on the reservation. Eighty-five percent of borrowers had never had a checking or savings account; 75 percent had never had a loan; and 95 percent had no business experience.

Today, Lakota has an over $4.4 million loan portfolio, has helped thousands of artists and aspiring entrepreneurs, and created over 1,000 jobs and hundreds of businesses.

One of the earliest "ecological economy" voices, E. F. Schumacher wrote the classic *Small Is Beautiful: Economics As If People Mattered* (1973). He knew that nature can only handle so much pollution, and the focus on rapid output of consumer products through advancing technology based on non-renewable fossil fuel was destructive and unsustainable.

Developed by The E.F. Schumacher Society *(schumacher.org.uk)*, the Self-Help Association for a Regional Economy (SHARE) is a nonprofit membership organization that partners with local banks to offer microcredit loans at manageable interest rates to businesses often considered "high risk" by traditional lenders. The bank makes the loans and handles the accounting, but the lending decisions, based on a unique set of social, ecological, and financial criteria established by SHARE, are made by the community of depositors, thus increasing their sense of responsibility.

A multi-program organization, the Society uses Community Land Trusts to foster "common land ownership ... that removes land from the speculative market." The Society also developed a local currency called BerkShares as a "tool for community empowerment, enabling merchants and consumers to plant the seeds for an alternative economic future for their communities."

Similar experiments with local currencies are popping up across the planet. Local—or community—currencies have social, economic, and environmental advantages. BerkShares claims that the currency helps area businesses connect with their customers but also strengthens the regional economy by favoring locals and keeps the money circulating within the community.

In the United Kingdom, "Transition Towns," which are seeking to use less oil, are exploring the environmental benefits of local currencies. According to Transition Town Totnes (the first Transition Town), "if we use our money for production and consumption closer to home, we're going to pay more attention to how those products are made, and the waste streams that result from them."

Transition Linlithgow, in West Lothian, Scotland, exists to "strengthen the local economy, reduce the cost of living and build resilience for a future with less cheap energy"

Choose Well[b]

"Economic competition, which today is being globalized, increasingly pits workers in each enterprise against workers in all enterprises, workers in each ethnic group against workers in all ethnic groups, and workers in each country against workers in all countries.

"Economic competition can only destroy social/environmental sustainability — never forge its links. We thus find ourselves oftentimes competing with the very people with whom we need to collaborate, which frequently leads to destructive conflicts over the way in which resources are used and who gets what, how much, and for how long.

"This need not be, however, because, we are not locked into any given circumstance. If the choices we make do not work, we can *always* choose to choose *again*. The final question will always be: How shall we choose?"

— Chris Maser

Earth Economy Tips

- Consume wisely:
  - Pay cash and shop at local independent stores; buy locally made/grown items
  - Join a cooperative and shop there; support worker-owned companies
  - Favor companies in the *National Green Pages*, and get tips from Green America
  - Ask for Fair Trade, FSC, and other third-party certified products
- Support or start a local BALLE affiliate network
- Support local currency, timeshares, barter, and other local economic innovations
- Use a local community development bank or credit union
- See greenwashing? Report it to the FTC and post on the Greenwashing Index at the University of Oregon
- Invest well:
  - Become a micro-lending investor
  - See Green America's *Investing in Communities* guide
  - Find a mutual fund through the Social Investment Forum's directory
- Get longer lists of tips from:
  - *Global to Local: What You Can Do*—ifg.org
  - *YES!* Magazine
  - *Green America—greenamerica.org*
  - *Better World Handbook*
  - *The Small-Mart Revolution*

EXPLORE & ENGAGE

The term "economics" strikes many people as dry and forbidding. The dominant culture's emphasis on money and profit has created deep inequities and a warped economic structure. Many feel chronic fear and scarcity. Questioning our assumptions around money allows for a different experience of wealth to emerge, one that is not dependent on a quantity of money in the bank. Innovative models based on localization and sustainability are now being explored and implemented throughout the world.

1. **Like the lyric, "let it begin with me," economic change begins with each of our own lifestyles and our relationship to money and wealth.**

> The question is not whether we can afford to invest in every child; it is whether we can afford not to.
>
> *— Marian Wright Edelman*

   Sit with your feelings about your personal economic well-being. What factors play a role in your emotions? Where would you place yourself on the rich/poor spectrum? Why? Who do you compare yourself to?

   Lynne Twist, founder of The Soul of Money Institute, suggests that no one is "poor," although many people live in impoverished circumstances. She says that labeling anyone "poor" makes it harder to value that individual's gifts. Discuss the implications of the "rich" and "poor" designations and how you feel about those terms.

   Explore new paradigms about money and wealth; watch one or both of these videos: Vicki Robin on "Your Money or Your Life" 👁 Lynne Twist on "Creating a Future For All of Us." 👁

   How might you reallocate your purchasing and investing to better align with supporting people, community, and the web of life? Decide on one small change you will make to create more alignment. (*Example: only buy coffee when you have your reusable cup with you; shop twice a month at a local farmers' market.*)

   Discuss some ways that you, your family, and your circle of influence can shift economic power to your local community.

2. **A regenerative society is a flourishing society. (p.84)**

   Recognize that we are living on Spaceship Earth. Keeping that reality in mind, what rules or principles would you adopt or create for a regenerative society? (*See the New Rules as a reference, p.84*)

➡ Choose and watch one of these movies that offer a glimpse at where our planet is currently headed: *Idiocracy* or *WALL-E*. What did you learn that expanded your perspective?

3. The current world economic model has been labeled "an empire," "the military industrial complex," and "economic globalization."

💡 What costs of the current dominant model of economics are of most concern to you? List at least a dozen ways institutions and individuals have been affected.

♥ Group Exercise: Visualizing an Earth Community (p.85)

- Play meditative, indigenous flute, or shamanic drum music.

- Choose a facilitator to set the stage—invite participants to relax, center themselves, and become present. Ask everyone to close their eyes and use their breath in order to drop from their thinking mind into their oneness heart. Have them feel earth energy to enter through their feet and up their spine.

- Invite participants to travel through time to a place where a peaceful community is serving the needs of the people and the planet in a sustainable and flourishing way. Ask them to imagine walking around this place, noticing how people are learning (pause) working (pause) eating (pause) playing (pause) and celebrating. Allow 5- 7 minutes.

- Then, have participants travel back to the current time with the experiential remembrance of what they encountered.

- Without talking, allow 2-4 minutes for each person to jot down notes about what they saw and felt.

- In groups of 3-5 people, have participants share what they experienced.

4. Global financial inequity affects everyone, whether in an arid region of Africa or in a city penthouse.

💡 Take in the following statistics:

- The four richest people in the world have more wealth than the billion poorest.

- The fortunes of the 792 billionaires increase annually by 300-400 billion.

- Global spending on armaments is a trillion dollars a year.

📖 Search the Internet for a way to comprehend what a "billion" or "trillion" actually means. Find an analogy or illustration that helps to make these numbers comprehensible and real.

📖 Spend three minutes watching a video that presents the earth's peoples as if they were a village of 100 inhabitants. 👁 What statistics help you connect with this reality?

📖 A movement that started in Great Britain, called "Uncut," has been taking action to bring attention to trans-nationals which make huge profits and are not paying taxes. Watch an Uncut video. 👁

➡ Share what you learn about this movement on social media or through email.

In a world of increasing inequality, the legitimacy of institutions that give precedence to the property rights of 'the Haves' over the human rights of 'the Have Nots' is inevitably called into serious question.

—*David Korten*

5. WTO, IMF, and NAFTA were designed to help trade and support economic recovery.

📖 On p.86 some of the negative side effects of these organizations are mentioned. Explore specific activities of the WTO, IMF, and NAFTA that have been detrimental to global human rights and sustainability—what would you suggest as a more beneficial approach?

❓ Watch the documentary "This is What Democracy Looks Like." 👁 What similarities exist between the WTO riots and the Occupy Wall Street movement? How is this an effective means of combatting deregulation for big corporations?

📖 Read John Perkins book *Confessions of an Economic Hit Man* or watch a great two-minute YouTube video of him speaking about it. 👁

💡 What suggested actions (either on p.89 or others you identify) are you willing to ask your legislative representative to take?

➡ Write a letter to your representative. Send it. Find the contact information for your representative online. 👁 Send a copy of your letter to at least three other people and ask them to write a similar letter.

6. Many groups are "flying under the radar" and making a difference to the economic welfare of individuals.

💡 How does the Fair Trade system—and similar certifications—illustrate a new economic model? What are some of the most significant tangible and intangible ways both producers and purchasers can benefit? Have you ever encountered your own resistance to paying more for Fair Trade? What choice did you make?

📖 Identify at least five groups that are using micro-lending or peer-to-peer lending to make a difference in the lives of girls and women. What are some of the reasons many groups focus on females? Find stories of recipients where the benefits have spread from individuals to family to community.

➡ Personal stories are powerful ways to allow us to feel connected and inspired. What story has moved you to take action? Find a creative way to share it (consider performing, writing, or singing).

7. Now that "environmental" and "eco" have generally favorable connotations, businesses are taking actions that range from progress to pretense.

💡 Many companies attempt to influence their stakeholders by using "green" or "socially responsible" terminology. Define each of the following terms in two sentences, including how the concept might be used: *biomimicry, cradle to cradle, Corporate Social Responsibility (CSR), Triple Bottom Line (TBL), Socially Responsible Investing (SRI).*

📖 Learn about companies that are perceived as leaders in some aspect of sustainability and/or responsibility. (*A few are mentioned on p.93.*) Explain why you think they do or do not deserve their reputation.

📖 Watch the movie *Waste=Food* 👁 or watch the late Ray Anderson of Interface Global in his TED Talk. 👁

♥ Read David Suzuki's epiphany about business being "the only institution on earth…[that can] lead mankind out of this mess." (p.90) Divide your group in half and debate the topic. One team gives arguments to support the statement; the other opposes it.

❓ Watch Janine Benyus' presentation on biomimicry. 👁 How might implementing such technological innovations reshape the way in which humans relate to the natural world?

❓ Discuss the idea of corporate "greenwashing." (p.94) What resources can you use to determine if a company or product is truly green or simply greenwashing?

➡ Watch this brief news clip about greenwashing and how to recognize it. 👁

➡ Use social media or a blog to alert your community to a company that is greenwashing. Contact the company and see how they defend their actions.

Real wealth is ideas plus energy.

— *Buckminster Fuller*

8. **Local and community economic solutions are emerging that are demonstrating financial, societal, and personal benefits.**

> You can make positive deposits in your own economy every day by reading and listening to powerful, positive, life-changing content and by associating with encouraging and hope-building people.
>
> — *Zig Ziglar*

♥ Imagine you live in a small town. What strategies or solutions would you recommend to your neighbors and officials to bolster your town's economic and community viability? What business and financial institutions would you like to attract or set up? What models would you point to in order to make your case? What food and energy approaches would contribute to economic viability? What statistics would you cite?

📖 Explore the important website BALLE, Business Alliance for Living Economies: *livingeconomies.org*. ◉

❓ What local businesses, social enterprises, and cooperatives do you support? What is difficult or inconvenient working with them or purchasing from them? What expected and unexpected rewards have you discovered?

9. **As individuals, our purchasing choices affect what products will continue to be produced and supplied; this is the principle behind the admonition "vote with your dollars." It is one of our strongest personal impacts in the world, for better or for worse.**

❓ View the documentary *No Logo*, a documentary about the power and pervasiveness of brands and the consequences to our world. How does branding influence the way we think and consume? See the shorter intro video ◉ or full version. ◉

➡ After viewing *The Story of Stuff*, ◉ identify one change you will make in your daily or weekly spending habits. What would happen if 50% of the population in your community made that same change?

📖 What sources will you utilize to identify businesses you might "vote for" because their business practices are in line with your values?

Storyofstuff.org

10. **We hear a great deal of discussion on mainstream news about the state of the economy.**

💡 How does the state of the economy affect you on a personal level? Which of the following emotional states have you experienced regarding economic trends: hopeful, anxious, empowered, apathetic, frustrated, curious, scared?

- What alternative news sources provide you with information about what is going on at the local and grassroots economic level that may not be covered elsewhere? Discuss and find out what sources others are using.

- What spiritual practices have you found helpful in maintaining a consciousness that is less buffeted by the money fear and anger often portrayed or incited by the media?

- Consider what your economic situation would be if you had been born in another country with limited educational opportunities, surviving on handcrafts, not knowing whether you will be able to purchase enough food from day to day to feed yourself and your family.

  - Reflect on the blessings and privileges of your present life situation and the opportunities that you have that so many do not.

- Make a commitment to make one beneficial change in your personal economy—how you work your finances—or your attitude about it.

Big Stock Photo

Part Six: COMMUNITY

Living Well Together: Growing and Nurturing Strong Communities

A too highly developed
individualism can lead to a
debilitating sense of isolation so
that you can be lonely
and lost in a crowd...
Ubuntu speaks to the essence
of being human.
The solitary individual is,
in our understanding,
a contradiction in terms.

— Archbishop Desmond Tutu

Every community can be part of the world-wide sustainability movement. Take Lexington, Kentucky—it may not be the first place that comes to mind as being on the environmental cutting edge, but for almost a decade Lexington has pursued and developed a host of initiatives designed to bolster the strength of its community while improving the health of the planet.

Many of these initiatives are the work of a non-profit called the Sustainability Communities Network (sustainlex.org), which brings together faith, education, government, business, and community-based efforts on projects that include a community garden, a youth green corps, and a university sustainability task force.

What has emerged in Lexington exemplifies what is possible and what is, in fact, emerging in communities everywhere. The Sustainability Communities Network is guided by four principles:

- environmental stewardship (the foundation)
- economic prosperity
- community empowerment
- social equity

These provide the framework for bringing together previously disparate elements of what Paul Hawken calls "the movement that doesn't yet know itself as a movement."

The challenge, for communities like Lexington and for the multitude of environmental initiatives taking root around the world, is how to connect the dots—how to transform our disconnected efforts into the alchemy of integrated systems.

"You need other people in order to be. You need other beings in order to be…you also need sunshine, river, air, trees, birds, elephants, and so on. So it is impossible to be by yourself, alone. You have to ' inter-be' with everyone and everything else."

—*Thich Nhat Hanh, Buddhist monk and peace activist*

"Never doubt that a small group of thoughtful, committed people can change the world. Indeed, it is the only thing that ever has."

—*Margaret Mead, cultural anthropologist*

BRINGING THE WISDOM OF THE COMMUNITY TOGETHER

"Social process may be conceived as either the opposing battle of desires with the victory of one over the other, or the integrating of desires. The former means non-freedom for both sides, the defeated bound to the victor, the victor bound to the false situation thus created — both bound. The latter means a freeing for both sides and increased total power—increased capacity in the world." (Mary Parker Follett).

How do we begin to come together and create a future based on the emerging wisdom of interconnectedness and sufficiency? How do we organize ourselves *in* and *as* community around issues of sustainability? How do we do so in a fair and inclusive way?

Dozens of initiatives foster working together toward a sustainable future, in "community"—a word that could be an abbreviation of *coming together in unity*. Most tools for embodying community focus on interdependence — tools of co-creation, co-intelligence, building united vision, and healing through communication.

Skills for integrating voices, desires and needs are foundational to growing successful community-based initiatives. These include communicating clearly, sharing power, collaboration, and honoring diversity of culture, background and skills.

- Some examples of powerful models for building community through enhancing relationships:

- Tom Atlee's *The Tao of Democracy* provides an overview of many excellent exercises, frameworks, and resources to develop such skills.

- Nonviolent Communication *(cnvc.org)* is a simple yet profound "language" that offers a needs-based way of transforming what is possible through communication.

- Relational Presence *(speakingcircles.com)* is another practice for building human connections through the simple power of awareness and intention.

- Sociocracy, or Dynamic Governance *(governancealive.com/dynamic-governance)*, is a consensus-

Aboriginal village council meeting in Austrailia

Community Meeting/AusAID

I'd like to make a deposit

based process for self-organizing groups.

- Other group process approaches worth exploring include Open Space Technology *(en.wikipedia. org/wiki/Open_Space_Technology)* and Dynamic Facilitation *(tobe.net)*.

How do we link groups of people who want to embrace more sustainable lifestyles? It's as simple as starting a series of conversations, as shown in the initiatives highlighted in this chapter. Many effective models are emerging for rapidly diffusing the knowledge, skills, tools, and support needed for broad-scale, systemic community change. Each uses some version of a time-honored format—self-organizing circles of people joining forces to build momentum and to create ripples of enlightened action.

- The book you're reading, for example, can become a powerful tool for engagement in community through the Sustainable World Coalition's "Engagement Circles" program *(swcoalition.org/ programs/engagement-circles)*, which offers free facilitation resources to anyone wishing to start a circle to deepen the learning from this book and get into action.

- More than 140,000 people have participated in Northwest Earth Institute *(nwei.org/discussion_ courses)* discussion courses, which foster education and engagement in the context of a small community.

- Empowerment Institute *(empowermentinstitute. net)* offers team-oriented programs for creating sustainable lifestyles and communities.

- Be the Change Circles *(bethechangeearthalliance. org/circles)* offer study-action-support groups with a variety of curricula, as well as periodic symposia and quarterly gatherings. This is a great way to stay engaged and continue to educate yourself on the issues so you can become an effective agent of change.

- For those wishing to clarify their own unique role in the shift to a thriving planetary civilization, What's Your Tree? *(whatsyourtree.it)* circle-courses support participants in finding and acting on their purpose, passion, and personal power within a growing network.

- The Center for Partnership Studies *(partnership-way.org)* offers tools to create groups committed to facilitating the emerging shift to a "Partnership" as opposed to "Dominator" society.

- The Real Wealth Community Project *(realwealth. pdx)* and the Caring Economy Leadership Program *(caringeconomy.org/CELP)* are training scores of leaders to engage their communities in lively discussions about creating an economic system that values people who do the work of caring for others and a sustainable planet.

- The Transition Network *(transitionnetwork.org)* has designed a comprehensive, systemic model for communities to organize on a small scale in response to global environmental and social problems. These "Transition Initiatives" create and implement plans to build sustainable and resilient communities—and any community can sign on. These plans include scaling back fossil fuel consumption, but Transition Initiatives also include projects in education, housing, waste reduction, and food production. Links to resources, toolkits and a growing network of "Transition Towns" in more than 30 countries are available at transitionnetwork.org.

TRANSITION TOWNS

The Transition Network has identified a series of stages in the process of transitioning to a self-reliant, sustainable geographic community. While all Transition Towns share certain things in common, each group determines what works best for their community. Here are the network's five stages with a sampling of "ingredients" of each stage:

1. **Starting out**

    This first stage transforms a mere idea or aspiration of "Transition" to a consensual vision, grounded in the intentions of a community
    - Coming together as a group
    - Visioning
    - Awareness raising
    - Building partnerships

2. **Deepening**

    The Transition initiative builds momentum, and practical projects start to emerge. A plan is designed to sustain the organization and deepen its work, while broadening community engagement
    - Skill building for local resilience
    - Development of local food initiatives
    - Education for the transition process

3. **Connecting**

    The scale of a proper response to peak oil and climate change has been likened to the preparations for World War II. Every aspect of our lives needs to change, in a coordinated and effective way.
    - Forming networks of Transition initiatives
    - Working with local businesses
    - Engaging young people
    - Pausing for reflection

4. **Building**

    Ultimately, transition groups aim to localize their community's economy. They move from running small community projects to acting on a larger scale with greater impact. New skills and ways of thinking lead Transition projects to become social enterprises, such as developers, banks, and utility companies.
    - Social enterprise and entrepreneurship
    - Scaling up
    - Strategic local infrastructure
    - Intermediate technologies

5. **Daring to dream**

    The old saying "Think globally, act locally" is still relevant. The ingredients in this section imagine elevating Transition thinking from the local to the national stage—imagining every community with a vibrant Transition initiative, setting up food networks, energy companies, resource and skill-sharing networks, and catalyzing a new culture of social enterprise.
    - Policies for Transition
    - A learning network
    - Investing in Transition
    - From TransitionNetwork.org

GROWING COMMUNITIES
Land Use and Local Government

"When we see the land as community to which we belong, we may begin to use it with love and respect."
    —*Aldo Leopold, author, A Sand County Almanac*

Re-envisioning how we relate to and use the land is fundamental to creating a sustainable future. From there, we must learn how to work with local and state governments to influence land-use policies and practices. That may sound daunting, but plenty of resources are available for working successfully with local government.

Take the Institute for Sustainable Communities *(iscvt.org)*, for example. This nonprofit helped the people of Moss Point, Mississippi and other Gulf Coast communities recover from the devastation of Hurricane Katrina by working with city officials to create a neighborhood advisory committee to assist in the city-rebuilding planning process. The skills and connections developed through that process then helped residents respond more quickly to the catastrophic Deepwater Horizon oil spill.

Similar efforts are being launched everywhere:

For planning, building, and development efforts that meet sustainability criteria, check out The Urban Land Institute *(uli.org)* and the Project for Public Spaces *(pps.org)*.

Smart Communities *(smartcommunities.org)* offers support and success stories both for renovating existing structures and for building new ones.

Partners for Livable Communities *(livable.org)* has an Aging in Place initiative, developed to help US communities become places that are good to grow old in.

Whether you are a landowner interested in protecting property from future development or a citizen wanting to mobilize your community to protect undeveloped land, the Land Trust Alliance *(landtrustalliance.org)* can help. One successful example: western North Carolina's Blue Ridge Forever initiative *(blueridgeforever.info)*, in which 13 organizations joined together to protect 50,000 acres of land and water through agreements and acquisitions.

Many communities are implementing sustainable economic alternatives, such as Time Banks *(timebanks.org)*, which track and exchange credits for time spent doing anything for someone else, and the similarly-purposed Local Exchange Trading Systems (LETS).

GreenAmerica *(greenamerica.org)* offers useful tips for individuals and communities trying to be sustainable and thrifty, such as starting a free store and caring for elders and children cooperatively.

The Business Alliance for Local Living Economies *(bealocalist.org)* is building a vast international network of communities, bringing together small business leaders, investors, entrepreneurs, development professionals, government officials, social innovators, and community leaders. Their Living Economy Principles are an excellent roadmap for creating sustainable communities.

Thinkstock

> *"A person's heart away from nature becomes hard...lack of respect for growing living things soon leads to lack of respect for humans, too."*
>
> —Lakota proverb

Sometimes community planning starts literally from the ground up, as in Arcosanti, an experimental "intentional community" in Arizona's high desert that melds architecture with ecology. "Intentional community" is an inclusive term for ecovillages, co-housing arrangements, residential land trusts, communes, urban housing cooperatives, and other planned communities where people live together with a common vision.

The website ic.org is an excellent resource for such communities at every stage of planning and formation. It also offers's *Communities* magazine, whose mission is "creating and enhancing community in the workplace, in nonprofit or activist organizations, and in neighborhoods."

See the sidebar for three model intentional communities. But there are also communities across the country coming together in less formal ways, introducing neighbors to each other—who then begin to share meals, pet and childcare, garden produce and plants, tools and other goods and services. These informal groups also make community organizing easier when needed to achieve community goals like land-use planning, safety, emergency planning, and improved transportation (e.g. bike-safe streets and crosswalks).

Three Model Communities

Sirius—*Shutesbury, Massachusetts*. With organic gardens, renewable energy, and vegetable oil-powered vehicles, the residents of Sirius have incorporated ecology and sustainability into every aspect of their community. Founded in 1978 on 90 acres of land, Sirius is modeled after Scotland's Findhorn community and is now home to about 30 residents who govern by consensus.

Buildings rely on sun and wind for electricity and are made from non-toxic materials. On-site gardens and greenhouses supply most of the community's shared meals. What little food they do not grow is purchased in bulk, and all scraps are composted. For non-residents, Sirius also offers classes in permaculture, off-grid energy, and organic gardening.

"Ecology comes from a deep inner sense of the sacredness of life. We want people to have a deep experience so they will alter their life patterns."[1]

Hammarby Sjöstad—*Stockholm, Sweden*. Author Bill McKibben writes that in this community residents "live half again as lightly as the average Swede, who is already among the most ecologically minded citizens of the developed world … [consuming at] the level calculated to be sustainable for all the world's seven billion humans."[2]

This former industrial brownfield was developed to be an environmental role model. Garbage is almost eliminated, with combustibles burned to produce heat and electricity and food scraps composted.

Wastewater has multiple lives: the sewage treatment plant separates liquids and sludge; sludge produces biogas to fuel kitchen stoves and buses, as well as fertilizer for farms. Treated wastewater goes into radiators to heat apartments. As this water cools, it is used to cool office computer rooms and grocery store coolers.[3] The City Project Manager says that "75% of Hammarby's sustainability is integrated into buildings and infrastructure—the remaining 25% is up to the residents themselves." The Hammarby model has been exported to Russia, England, and China.[4]

Los Angeles Ecovillage—*Los Angeles, California*. Some 500 people live in this two-block working-class neighborhood near downtown LA. After the 1992 civil unrest in Los Angeles, several residents decided to transform their own troubled neighborhood.

The first three years of the project involved building trust and a sense of safety, planting trees and small gardens, and acquiring property. The first purchase was a 40-unit apartment building purchased with the aid of a community revolving loan fund. Substantial renovations were completed, mostly by residents trained to do this work using sustainable and recycled materials. One unit is a "common house" for gatherings, and another is for bicycle parking. Residents without cars get a small rent discount.

These were among the first LA apartments to set up an extensive recycling program, influencing neighbors to recycle as well. To encourage re-use, residents put items they no longer needed on a "free" table. A self-help bicycle repair shop called The Bicycle Kitchen was started in the kitchen of one of the apartment units.

The Food Lobby, an organic food-buying coop, minimizes waste through bulk produce and grocery purchases. Composting is part of everyday life in the LA Ecovillage, and the neighborhood now showcases several water-harvesting permaculture gardens. A variety of small business start-ups provide a livelihood for several neighbors, including fair-trade coffee, vegan chocolates, and custom bike-building companies.

Food and meals

What could be more fundamental than the food we eat? With resounding effects on our health and the planet's health, industrial agriculture is failing the entire world by contributing to climate change, pollution, inhumane treatment of animals, habitat destruction, hazardous low-wage working conditions and human illness. For more information, visit the Union of Concerned Scientists' website *(ucsusa.org/food_and_agriculture)*.

The landscape of "Big Agriculture" may look pretty bleak, but each of us has opportunities every day to make healthier choices for ourselves and the environment.

A great way to start is by eating locally, whether that means visiting your local farmers market or, even better, signing up for a Community Supported Agriculture (CSA) membership, through which you pay upfront for boxes of fresh produce delivered throughout the harvest season. Check out localharvest.org to find CSA's near you. Also, coopdirectory.org offers a state-by-state directory of co-ops that sell healthy foods.

Of course, the ultimate in eating locally is to grow your own food. Even apartment dwellers can participate. To join, start, or help defend a community garden, The American Community Gardening Association *(communitygarden.org)* offers practical advice. Food Not Lawns *(foodnotlawns.com)* offers educational, organizational, and hands-on services to support transitioning from lawns to homegrown food, and Sharing Backyards links people with unused yard space with those looking for a place to grow food. *Local Food: How to Make It Happen in Your Community*, by Tamzin Pinkerton, is the first in a series of how-to books based on wisdom gained by Transition Towns around the world.

Larger-scale community efforts to make qual-

Creative Commons/Corey Templeton

ity food available across the social spectrum include Food Not Bombs *(foodnotbombs.net)*, an all-volunteer organization though which local groups recover food that would otherwise be discarded and then make free, fresh, hot, vegan and vegetarian meals served outside in public spaces.

Farm Fresh Choice, a program of Berkeley, California's Ecology Center *(ecologycenter.org)*, supports healthy local agriculture as well as the health of poorer citizens nearby. Fresh, organic fruit and veggies are purchased wholesale from growers at the Berkeley Farmers Market, then sold at wholesale prices to residents at markets set up near schools and childcare centers.

Re-imagining Sustainable Culture

Community-based arts and cultural organizations supported by the Ford Foundation are revitalizing neighborhoods. They partner with traditional community development groups to build common vision, foster tolerance and respect, and boost economic prospects in rapidly changing underserved neighborhoods.

A key goal of Partners for Livable Communities' Shifting Sands Initiative is to accumulate best practices that highlight a new role for arts and culture groups in community development, through programs supporting social integration across race and class, upward economic mobility, neighborhood identity, and civic engagement.

Beginning and Ending: Home birth with the aid of midwives and doulas has long been an option for healthy childbirth. thematrona.com has been "promoting quantum midwifery and undisturbed birth in the global village since 2001." The benefits of nursing infants are numerous. It's free, of course, and healthy, and it avoids supporting corporations that synthesize, package, ship, and advertise chemicals to vulnerable populations. Voices.org has information on the importance of breastfeeding.

Good models abound for cooperative childcare, but what about caring for people near the end of life? Far fewer cooperative eldercare models exist, though the need grows in the face of an expanding senior population. Green burial, using a biodegradable cas-

ket without embalming, is also gaining recognition as a final testimony to the impact we can choose not to have on the environment.

Taking Care: Based on the psychology of interdependence, Gentle Teaching International *(gentle-teaching.sharevision.ca)* trains caregivers and companions of those who have inherent vulnerabilities, such as extreme poverty, homelessness, mental disability or illness, focusing on being kind, nurturing, and loving.

Waste Not, Want Not: Among our most overlooked natural resources are the skills honed by our elders, especially those who lived through the Great Depression. Since the economic downturn of 2008, an increasing number of individuals and families are returning to the practices and perspectives of an older generation that conserved in every respect, cultivating ingenuity, resourcefulness, and commitment to community.

Sharing labor, food, child and elder care, along with enjoyment of simple (often free) pleasures, supports the well- being of individuals and families, while building a strong sense of community.

Sharing resources: Many communities have established "tool libraries" to encourage sharing rather than purchasing. (See for example the North Portland, Oregon, Tool Library: neptl.org. Extended groups of friends, neighbors and acquaintances have established plant and seed swaps that encourage gardening).

These activities can be initiated informally and may easily grow to serve ever-widening circles within your community. Many communities also now have arts, crafts and carpentry skill centers that offer classes, workshop space, and community interaction. These can also foster locally-based activities that support local artisans and businesses.

In some communities, households have begun to share services like garbage and recycling pick-up, among others—both saving money and building community connections.

Where to begin in your community? Organize a block party. This sounds simple but community-building often begins simply by getting to know your neighbors. Sharing food, music and conversation on a traffic-free street can be an ideal and relatively easy way to explore what other community activities might be possible.

GREEN BUILDING

A Green Village Begins with Green Homes

Imagine being outside in winter with holes in your pockets and rips in your garments. Your money and your body heat are leaking out with each passing hour. Inefficiently planned and insulated buildings squander precious resources in much the same way. We spend 90% of our lives inside buildings.[5] Accordingly, enhancing building performance—energy efficiency, indoor air quality, etc.—is a big step toward reducing overall energy consumption.

Much like organic bodies, buildings are coordinated systems that protect us and keep us warm. The future will see *smart buildings* responding so seamlessly to our sweat, our shivers, our need for hot water and other shelter-related details that we will regard our present-day programmable thermostats as we now do phonographs.

But right now we need those thermostats, along with every other technology available, from ancient (plugging the leak and facing the sun) to modern (solar panels and LED light bulbs) to combinations of both (geothermal applications with vast swaths of urban green roof "prairies"). We also need major investments—public and private—in commercial building retrofits.

Green building is a phrase with wings so wide it covers everything from the prosaic to the visionary to building entirely new communities from scratch.

"Green, or sustainable, building refers to an evolving set of construction practices that consider energy and resource use; environmental and site impact; product and building durability; and the impact of the built environment on occupant health and safety."[6] When take into consideration through every step of the design and construction process, a building's negative impact on people and planet is significantly reduced.

Several rating systems are in use for green buildings: Energy Star/Green Star *(energystar.gov)*, LEED (Leadership in Energy and Environmental Design) *(usgbc.org/leed),* and BREEAM (the UK's Building Research Establishment Environmental Assessment Method) *(bream.org).*

LEED's thorough checklist of green building components is often linked to US tax incentives, while LEED certification, which has several levels,

Green Roof at Walter Reed Community Center, Arlington, VA

raises property values and is increasingly coveted by residential and commercial real estate investors.

Perhaps the most rigorous of all green building certifications, the Living Building Challenge *(living-future.org/lbc)* requires that a building "generates all of its own energy with renewable nontoxic resources, captures and treats all of its water, and operates efficiently and for maximum beauty."

Shelter Stats

- 41% of US energy—7% of all energy used in the world—goes toward powering buildings. Much is wasted through poor insulation, leaky windows, poor construction technique, and inefficient lighting, heating, or cooling systems.[7]

- In the US, buildings account for 38% of all CO_2 emissions.[8]

- US residences, on average, are at least twice the size of typical homes in Europe and Japan, consuming 2.4 times the energy.[9]

Seven Steps to a Sustainable Building

- **Siting** and **orientation**: The greatest gifts for a high-performance building.[10]

- **Energy Efficiency:** Good insulation as well as low-energy lighting and appliances.

- **Natural Light, Heating, and Cooling:** Windows and skylights provide daytime light, while thermal mass, overhangs, and ventilation help occupants stay comfortable. Once installed, it's free energy.

- **Indoor Environmental Quality:** Balancing a tight building envelope against the need for fresh air. Many materials and products—like paint and particleboard—off-gas, so low or no-VOC (Volatile Organic Compounds)-emitting products should be used. (See The Healthy Building Network, healthybuilding.net, for product information.)

- **Recycled and Renewable materials:** Local materials should be used when possible. Also, more durable materials can increase the building's

lifespan.

- **Water usage reductions:** Hardy landscaping, low-flow devices, and simple conservation.

- **Regenerative/Adaptive** technologies that positively respond to the greater environment.

GREENING CITIES

The *New Yorker*'s David Owen dubs New York City "the greenest community in America" for reasons including its density, shared walls (easier to heat), and high use of public transportation.[11] With additional GHG-reducing initiatives like green roofs, more efficient buildings, and commercial retrofits (like the massive greening of the Empire State Building launched in 2009), our densest cities can all maximize their strengths. Places like Portland, Ore., Chicago, Austin, Seattle, and San Francisco — and even urban centers in more traditionally conservative states like Texas and North Carolina — are all "seeing green."

Greening roofs: Every city has hundreds of acres of flat, black tar roofs, creating a *heat-island effect,* which causes urban areas to get hotter than rural landscapes. Simply painting roofs with white or solar-reflective paint could go a long way, cutting energy use by up to 40% in summer months.[12]

When it comes to roofs, white is a good start, but green is even better. Green roofs—covered with vegetation, reduce storm water runoff, last 50-plus years (as opposed to 20 years for black tar), increase the value of any nearby property with a view, and have a long-term fiscal payoff. Green rooftops immediately reduce carbon footprint, reduce heat generation, and are up to 60% cooler than conventional roofs.

Green walls, or vertical plantings, are also starting to take root across America, reducing cooling costs and adding value.[13] Green walls inside buildings also improve indoor air quality.

Urban Infill and Suburban Retrofits

Roads, utilities, transportation: The infrastructure needed to build new suburban neighborhoods comes with a hefty price tag compared to building within

Compost in the making

cities—"urban infill." Suburban construction also has substantial repercussions for the environment, our cities and our health.

Building up urban environments conserves natural landscape while increasing the density of urban populations. That in turn supports walkability and public transit, reducing our dependence on cars. To see how walkable your neighborhood is, or to factor walkability into your relocation decisions, look up your Walk Score at *walkscore.com*.

Retrofitting our existing suburban neighborhoods to be less car-dependent will also play a large role in greening our cities. Goals for suburban retrofits would be, for example, to increase density, add mixed-use zoning, or repurpose vacant big-box stores.

Water and Sewage

Stewardship of our natural resources is the sine qua non of sustainable living. Conscientious communities establish practices like water harvesting, conservation, and management of solid waste and greywater.

The Greywater Guerrillas *(greywateraction.org)* are educators, designers, builders, and artists whose mission is to empower people to build sustainable water culture and infrastructure. Ecological sanita-

tion—i.e. the composting toilet—has made impressive improvements to health and cleanliness in rural regions. Soon those same methods can be implemented widely in urban environments. Greywater options include routing your used washing machine water, dishes water and shower water into your garden (make sure you are using bio-degradeable soap!).

Greening wastewater: Innovations are flowering in wastewater treatment technology. The Solid Immobilized Bio-Filter system, for example, eliminates some drawbacks of conventional treatment plants, costs less to start up and operate, and can save 90% of ongoing energy costs.

Reusable treated water is used for groundwater recharge as well as on- and off-site gardening irrigation opportunities.[14] New Energy Star toilets offer useful flush options, as do greywater recycling toilets and sinks.

Installing a 3000 gallon water tank to collect rainwater runoff from your home's roof can be used to water the garden as well as backup drinking water. Aerating faucets in sinks and tubs conserve water and extend its use. Encourage your family/community to take shorter showers—the energy savings alone can be considerable.

TRANSPORTATION

Rethinking How We Get Around

Heavily reliant upon individual automobile ownership and weakened by an aging infrastructure, a heavily indebted US now faces a host of transportation-oriented crises.

Scientists tell us we have little time before catastrophic climate change becomes inevitable and irreversible. Despite this urgency—and a short-term

---

How Can I Influence Policy and Still Have Time to Cook Dinner?

- **Group your representatives** in your contact list and lobby them! Email or call when you hear about green building initiatives. Write an email template for yourself: "I support xyz," plus your personal standard message, and just change it a bit for each initiative you write about.

- **Become a fan**. Most groups have Facebook pages. While you're looking at your friends' latest posts, click on a link from the US Green Building Council (*usgbc.org*) or Inhabitat.com. Sign up for their email newsletters.

- **Take a peek** at amazing futuristic designs! Explore for a few minutes a day, and before you know it you'll be a green building info whiz, ready to educate your friends, family, and community.

- **Get inspired**. Creative minds are sparking new ideas across the planet. Explore these ideas in a circle of like-minded friends to support each other in acting together.

---

decrease in annual vehicle-miles traveled due to gas price spikes—we are still witnessing an increase in the overall number of vehicles in use and a corresponding rise in vehicle miles traveled.

According to the US Environmental Protection Agency, the combustion of fossil fuels to transport people and goods is the second-largest source of CO_2 emissions.[15] Motor vehicles are also major contributors of carbon monoxide, nitrogen oxides, and volatile organic compounds, contributing mightily to ground-level ozone and smog.[16] Ozone and suspended particles cause and exacerbate respiratory ailments, which disproportionately afflict communities of color and lower income.

Back to the Hood

Sprawling development throughout the last half of the 20th century created transportation quagmires across the US. Changes at both the personal and policy level are critical as communities cast a keen eye toward eco-sensitive development, wise land-use planning and smart growth.

Individuals and families must find ways to stem dependence on private automobiles, such as shortening, sharing or even eliminating commutes (think telecommuting). Taxes at the pump can become revenue streams for public transportation systems: buses, light rail, cycling paths, walkable downtowns and communities. For intercity travel, the future belongs to high-speed trains.

Taking public transit, ridesharing, biking, or walking may lengthen commutes. But toward the mission of slowing global warming and the irreversible global damage it will wreak, the few extra minutes per day become time well spent.

Such lifestyle shifts accrue multiple benefits, including keeping money circulating in local communities, spending less of our lives on gridlocked highways, reducing fossil fuel emissions, and protecting the Earth in ways yet to be realized.

Sustainable Transportation Planning Recommendations: According to the Center for Transit-Oriented Development *(ctod.org)*, "smart growth" links transit systems to networks of walkable and bike-able streets and mixed-use retail and residential areas, including workplaces and affordable housing, It improves quality of life and reduces household transportation expenses. Any region can benefit from stable mixed-use—and mixed-income—neighborhoods that reduce environmental impacts and traffic congestion.

What You Can Do

- Turn off the engine rather than allow your car to idle. Engines emit three times as much exhaust when idling than when powering a vehicle.

- Try going car-free just one day a week for a month. You'll be surprised at the benefits. Many communities and organizations now offer active ride-share programs and organizations that encourage biking, including for families on their way to school. Find one near you. This can be a good way to meet neighbors and forge community connections.

- Start an anti-idling policy in your town. Many communities have already instituted no-idling policies around schools. The US EPA has a National Idle Reduction Campaign aimed at school buses (see epa.gov/cleanschoolbus/antiidling.htm.) Find out if this can be done in your community or extended to include other community areas.

Redesigning Urban Transport: What Works

Urbanization is one of the most significant modern-day phenomena, making more sustainable urban transport an urgent priority. Six out of 10 people will live in cities by 2030.[17]

The world's fast-growing cities are struggling to provide services to residents and cope with overburdened roads and rising road traffic injuries. City planners and citizen activists are finding that a blend of rail, bus, bicycle lanes, and pedestrian pathways offers the best set of options for mobility, cost effectiveness, and a healthy and pleasing environment.

Optimally, rail provides the hub with bus lines intersecting. For example, in Bogotá, Colombia, a highly successful bus rapid transit system, Trans-Milenio, uses special express lanes to move people quickly through the city.

This is being replicated in other countries.[18] Japan's legendary bullet trains have carried billions over the past 40 years without a single casualty, while China has 5,800 miles[19] of high-speed rail, serving about 1.33 million passengers per day in 2012.[20]

Carbon dioxide emissions from high-speed trains are one-third those of cars and a quarter those of airplanes. If the grid were powered by green electricity, emissions from trains would be zero.

London, Singapore, Stockholm, Milan, and other cities are reducing traffic congestion and air pollution by charging cars to enter the city. In London, vehicle traffic declined immediately. In the first year, bus ridership increased 38% while traffic delays decreased 30%.[21]

In Copenhagen, each day 500,000 people commute by bike to work or school. The Danish are now investing $47 million in building "bicycle superhighways" extending far into the suburbs. Elements include smooth, even surfaces free of leaves, ice, and snow with sufficient width for passing, "service stations" with air and tools along the routes, and "green wave" sections with timed stoplights (cycle 12 miles per hour and you hit green lights all the way).[22]

Even though the US government is trying to expand public transit coverage, it lags far behind many countries in diversifying transit options. The National Complete Streets Coalition (*smartgrowthamerica. org/complete-streets*) lobbies for streets that are also friendly to pedestrians and bicycles, so far establishing "complete street" policies in 14 states and 40 metropolitan areas.[23]

Greener Trucking: Since it will be with us for a while longer, trucking needs to be cleaned up. Heavy-duty, long-haul truck smokestacks emit 6% of the United States' carbon dioxide each year. Doubling the efficiency of trucks from 6.5 miles per gallon to 12.3 mpg could save 3.8 billion gallons of diesel annually. According to the Rocky Mountain Institute, this could be done with readily available technology, including auxiliary power units, more efficient wide-base tires, and improved aerodynamic mechanisms such as trailer side skirts.[24] And no idling!

Bicycles are the main mode of transportation in some parts of the world

Hybrids and Electric Vehicles: The Best for Now

Each gallon of gasoline emits an amazing 24 to 28 pounds of carbon dioxide.[25] While federal regulations require the fuel economy of cars and trucks to double on average by 2025, there's still a long way to go. Until more sustainable transportation systems are created, alternative vehicles are necessary—but not sufficient—to slow global warming.

While ample information on hybrid and hydrogen fuel-cell vehicles is readily available in the mainstream press, for depth of understanding look to The Union of Concerned Scientists' website comparing hybrid vehicles *(hybridcenter.org)*. Looking for the most efficient car to buy? Check out the American Council for an Energy Efficient Economy's Green Book *(greenercars.com)*.

Ridesharing

Ridesharing is gaining momentum, from companies and campuses to online postings on sites like Craigslist. GreenRide *(greenride.com)* now coordinates ridesharing, rail, and bicycle routes across the United States, Canada, and Europe. It has saved 31 million miles of driving and prevented 13,000 tons of carbon dioxide emissions.

The University of Florida in Gainesville, the nation's fifth-largest campus, was the first American university to implement the GreenRide program. The school initially experimented with its 17,000 employees, then extended it to its 46,000 students. GreenRide matches not only student class schedules, but also weekend and vacation destinations. The service includes an online calculator that determines how much money and pollution is saved.

GreenRide usage saved the university $1 million —the cost of a new parking structure—demonstrating that making sustainable choices can be both good for the planet and financially prudent.

Meanwhile, car sharing services like Zipcar allow members to make occasional car trips without owning their own vehicle. The result? Fewer cars on the road. One shared car replaces about 10 privately owned vehicles, according to researchers at the University of California, Berkeley.[26] Check out *carsharing.net* for resources near you.

What You Can Do

Sierra Club's green transportation (sierraclub.org/transportation) initiative promotes actions everyone can take to reduce their carbon footprint. The good news is that you can take many small steps on a daily basis to do your part in the fight against global warming:

- Drive less!

- If you're in the market for a new car, buy the most fuel-efficient vehicle you can. Better fuel economy = a better environment.

- Check tire pressure frequently and keep tires fully inflated; this can improve your fuel economy up to 10%. Also keep your car tuned up.

- Use a GPS—a navigational device can reduce miles traveled up to 16%.

- Sell your car and join a car sharing company. See carsharing.net.

- Find out your car's optimal speed for fuel economy and set your cruise control.

- Choose an efficient route for your errand-running and combine errands to avoid multiple short trips. If you plan to make multiple stops at a shopping center, park your car in the middle and walk to your individual destinations. It all adds up—and cars emit more pollution in the first mile than in the next 10 miles!

- Roughly 44% of car trips taken are fewer than two miles. Burn calories instead of gasoline — walk or ride a bike. Commit to taking public transit, walking, or riding a bicycle at least one day a week.

- Check into the possibility of telecommuting for work, even one day a week.

- Carpool with co-workers. If a daily carpool won't work, try one or two days per week.

- Actively support public transportation in your community. Increased use and demand for public transportation can improve the level of service.

- Contact your federal legislators to allocate more funds for public transit than for highway construction.

- Express your support for raising taxes on gasoline as a stimulus to get people into alternative modes of transportation.

- If your public transit system excludes bikes from rush hour, write or call and ask legislators to accommodate bicycles. This is when the most automobile trips and pollution could be saved!

- The Surface Transportation Policy Project (transact.org) promotes "location-efficient incentives"— targeted subsidies to employers who locate in transit-accessible places, and sprawl-prevention measures so people who cannot afford a car or choose not to own one can still access jobs.

THE EARTH CHARTER

The Earth Charter (*earthcharterinaction.org*) is one of the most unique documents in human history, created through a series of consensus-based dialogues, with thousands of people participating from all over the world over the course of more than a decade. Now endorsed by more than 4,500 organizations and governments, it inspires shared responsibility "to bring forth a sustainable global society founded on respect for nature, universal human rights, economic justice, and a culture of peace."

The Earth Charter Initiative promotes living by example with the help of The Action Guidelines, which suggest a resourceful focus on root causes of problems.

Principles

I. RESPECT AND CARE FOR THE COMMUNITY OF LIFE

1. Respect Earth and life in all its diversity.
2. Care for the community of life with understanding, compassion, and love.
3. Build democratic societies that are just, participatory, sustainable, and peaceful.
4. Secure Earth's bounty and beauty for present and future generations.

II. ECOLOGICAL INTEGRITY

5. Protect and restore the integrity of Earth's ecological systems, with special concern for biological diversity and the natural processes that sustain life.
6. Prevent harm as the best method of environmental protection and, when knowledge is limited, apply a precautionary approach.
7. Adopt patterns of production, consumption, and reproduction that safeguard Earth's regenerative capacities, human rights, and community well being.
8. Advance the study of ecological sustainability and promote the open exchange and wide application of the knowledge acquired.

III. SOCIAL AND ECONOMIC JUSTICE

9. Eradicate poverty as an ethical, social, and environmental imperative.
10. Uphold the right of all, without discrimination, to a natural and social environment supportive of human dignity, bodily health, and spiritual well being, with special attention to the rights of indigenous peoples and minorities.
11. Ensure that economic activities and institutions at all levels promote human development in an equitable and sustainable manner.
12. Affirm gender equality and equity as prerequisites to sustainable development and ensure universal access to education, health care, and economic opportunity.

IV. DEMOCRACY, NONVIOLENCE, AND PEACE

13. Strengthen democratic institutions at all levels, and provide transparency and accountability in governance, inclusive participation in decision-making, and access to justice.
14. Integrate into formal education and lifelong learning the knowledge, values, and skills needed for a sustainable way of life.
15. Treat all living beings with respect and consideration.
16. Promote a culture of tolerance, nonviolence, and peace.

"Let ours be a time remembered for the awakening of a new reverence for life, the firm resolve to achieve sustainability, the quickening of the struggle for justice and peace, and the joyful celebration of life."

—from the last paragraph of The Earth Charter

EXPLORE & ENGAGE

When a culture forgets the importance of community, feelings of separation, scarcity, and loneliness increase. When we come from connectedness, creativity and innovation blossom—collaboration is the key to finding the solutions we seek.

1. Community is essential to the well-being of each of us and to creating a world that works for all.

➡ Look into *The Spark* documentary ◉ about sustainable farming/community building in Katrina ravaged part of New Orleans.

💡 "Community" suggests we are meant to live with others and experience a sense of self as part of the whole. How does this idea resonate with you?

❓ Think about how you participate in various communities. What are the conditions, requirements, or expectations for participation? In what ways would you like to be more or less engaged in each of your communities? What new communities call to you? How might you initiate participation in one of them?

➡ Write a note of appreciation to the members of at least one of your communities.

➡ Watch the movie *Pay It Forward* or watch the trailer ◉ so you understand the title and assignment given in the film. Accept the assignment, and implement your version of *Pay It Forward* for a specific period of time.

💡 How might the concept of "paying it forward" change your communities if enough people adopted this practice?

2. Consider Thich Nhat Han's quote on inter-being: *"You need other people and other beings in order to be."* (p.109)

♥ Group Exercise: 5-minute Meditation on Inter-Being

• Have a volunteer read the following aloud, slowly and with feeling,

Sometimes our flame goes out, but is blown again into instant flame by an encounter with another human being.

— Albert Schweitzer

We are confronted with insurmountable opportunities.

— Pogo

with pauses between sentences:

Take three deep breaths.

Remember something you ate or drank for breakfast this morning.

Imagine the journey that brought it into existence—originating as an idea, then taking form.

Who touched this food with their hands? Who invested their hearts?

How did the sun, rain, and soil offer their contribution?

Imagine the families who labored— growing, harvesting, and getting it to you.

Picture the journey: the roads, the riverways, or the oceans along which it traveled.

Consider those who refined, packaged, transported, sold, or played a role in bringing this food to your body.

What parts of the earth and others still live in you now?

Now imagine all connected by a light or energy web.

Feel into this inter-being.

- Get with a partner and share your experience.

3. Connecting powerfully and vulnerably catalyzes growth and impels action.

📖 From the many community-building organizations or initiatives described, choose one that inspires you. Research it. (*Examples: NWEI discussion courses, Be The Change circles, Global Action Plan, Real Wealth Community Project, Transition Towns, Utne*)

❓ Create a two minute "elevator speech" about this organization and present it to the group. While each speaker is presenting, have other participants practice listening deeply in order to identify groups that resonate with them.

➡ Break into small groups according to which groups attracted you. (*Example: three of you may have resonated with Transition Towns.*) Choose an action step you can take, either individually or as a group, in the coming week. (*Reading several pages of a website, calling to see if there is a local group or event you might attend*) Record your specific commitment on your calendar. Agree on how you will hold each other accountable. Share your plan with the greater group.

💡 Notice how asking for support with accountability influences your impetus to take action.

❓ Discuss your findings and celebrate any successes at the following group meeting. Talk about potential next steps.

In the long history of humankind (and animal kind, too) those who learned to collaborate and improvise most effectively have prevailed.

—*Charles Darwin*

Global Exchange

4. Conventional building practices have significantly contributed to environmental degradation and species extinction.

💡 What statistics about shelter in the US (p.117) most surprised you?

➡ This TED talk provides a short and funny explanation of what parts of building green are overlooked, and how it compares to conventional home building methods. 👁

➡ This video shows how building green can improve the workplace environment and add beauty to a community. 👁

📖 Choose a building where you spend a lot of time. What correlations exist between the way the property was developed and built, and the environmental degradation/species habitat impact? *(Building materials, native vegetation and animal habitat, the construction design and process)*

❓ Break into small groups and discuss a new set of building conventions that would turn development into part of the solution rather than perpetuating problems. What policies and regulations would support this?

5. Stories about community shape our attitudes and behavior and tell us where we belong.

❓ What are some of the stories you were told growing up that communicated who you were, what you were part of, and who you could trust? What messages did you receive about your family, relatives, any religious group, or your neighborhood, town, or nation? Discuss with a partner or in the group.

💡 What are some stories about community that have inspired you?

➡ Commit to sharing one of these stories with someone outside the group within the next week.

💡 What are the key messages about community and belonging that mass media and advertising communicate?

6. The "greening" of cities has multiple benefits.

❓ Discuss the concepts and benefits of green rooftops, white rooftops, and water harvesting (p.117). Watch Pam Warhurst's TED talk on a community-based movement to promote urban agriculture in the town of Todmorden. 👁 Could this idea be successfully implemented elsewhere?

Spiritual practice is really about weaving a network of good relationships.

—Dhyani Ywahoo

Natalie Maynor

📖 Investigate and share how your own city or town is becoming more green. What initiatives, regulations, or laws affect (either support or hinder) your town's efforts to move in that direction?

7. In the US, transportation accounts for 29% of energy consumption.

⑦ What are three transportation issues that drive the use of fossil fuels beyond what the earth can sustain? Discuss these and suggest solutions.

➡ For one week, keep a journal of your travel. Track both the miles and hours you spend on a plane, train, bus, car, bike, foot…even astral travel! Identify how much time you spend alone in a car.

➡ List steps you could take to reduce fuel consumption. In front of the group, commit to something you will do to reduce your transportation carbon footprint. *(See p.121 for the Sierra Club's list of "What You Can Do.")*

Big Stock Photo

8. Your community starts with your living situation.

➡ Describe your current living situation and the type of building in which you live. Do you live alone, or with other people? What do you appreciate about where you live and who you live with? What do you not like? How much community do you experience with your current housing arrangement?

💡 Imagine your ideal living situation. How important is it for your home to reflect your environmental and community values? How do you weigh various factors and prioritize?

📖 Watch this brief video about a working ecovillage near Mt. Fuji, Japan. ◉ Research co-housing and intentional communities in your area. Take a field trip to one of these, and invite friends to go along. Share what you learn.

9. At times, the "bad news" about our current situation on earth can be disheartening.

Watch a six minute speech by Paul Hawken on the current global movement that has no name. ◉ It offers potent hope at this time in history.

After watching, discuss the following questions:

• What gives you hope or inspiration after watching this video?

• How is the movement he describes different from others?

Hope is not a feeling of certainty that everything ends well. Hope is a feeling that life and work have meaning.

— *Václav Havel*

Part Seven: GETTING PERSONAL
Our Choices Matter: Creating a Sustainable Future with Our Daily Actions

The time of the lone wolf is over.
Gather yourselves! Banish the word
"struggle" from your attitude
and your vocabulary.
All that we do now must be done
in a sacred manner and in celebration.
We are the ones we've been waiting for.

—The Elders, Oraibi,
Arizona, Hopi Nation

FOOD

We Are What We Eat

How and what we eat impacts not only the health of our bodies, but the health of the planet as well.

Food is closely linked to the most pressing sustainability issues we face: pollution, population, transportation, energy, social justice, economics, animal welfare, and more.

Developed over the past 50-plus years, the mass production of food has created catastrophic impacts —from environmental toxins to massive transportation-related carbon emissions, from biodiversity loss to environmental injustice. The processed foods created by these mass-production techniques are far less nutritious than fresh foods. Their consumption adds another immense disconnect between human life and the rest of nature.

With each dollar spent on food and drink, we

vote for the system that produces what we buy, whether it benefits our own health and that of the environment or wreaks havoc on both. By learning about what we eat, we can make informed and critical personal choices; we can vote consciously for a life-nourishing food system.

Books such as *Diet for a Small Planet* (Frances Moore Lappé), *Diet for a New America* (John Robbins), *Fast Food Nation* (Eric Schlosser), *Blessing the Hands that Feed Us* (Vicki Robin), *The Omnivore's Dilemma*, and *In Defense of Food* (Michael Pollan) have raised awareness about the personal and planetary impact of each food choice we make. At the pinnacle of healthy food choices sits the concept of a locally-sourced food supply.

Yes, a little short-term convenience may be sacrificed when you eat consciously. But once you realize its true cost, cheap food no longer seems inexpensive. The investment in thoughtfully chosen, quality nourishment becomes not an indulgence or a luxury but a form of personal and planetary health insurance.

A World of Food and a World of Want

Hunger affects alarming numbers of people around the world, especially children. Each day, about 16,000 children die due to hunger-related causes. That's one child every five seconds. An estimated 842 million people worldwide are hungry, and more are chronically malnourished. Current exponential population growth and the lack of sustainable food strategies

Young girl selling chicle (chewing gum) in Chiapa de Corzo, Chiapas, Mexico

will only prolong issues of world hunger, malnourishment, and starvation. The number of the world's hungry is projected to reach 1.2 billion by 2025.[1]

The Food Sovereignty movement champions the rights of communities and countries to decide on agricultural and land-use policies appropriate to their unique circumstances (see Food Sovereignty sidebar). It stresses the universal right to sufficient, healthy food and rejects the idea that food is a commodity to be exploited.

Food Sovereignty is a proactive approach to achieving world and local food security and ending control of food by profit-focused corporations demonstrating little interest in a community's long-term quality of life.

What's the Trouble with 'Big Food'?

On average, US produce is shipped 1500 miles—more if you include produce imported from other countries and continents. Transporting food these great distances carries a double whammy: First, the nutritional value of imported produce diminishes each day it is in transit. Secondly, the fossil fuels used to transport food (not to mention those used in producing synthetic fertilizers to help food grow) contribute enormously to global warming and pollution.

Given the vast transport distances, what makes internationally produced food available at such cheap prices? Subsidized energy prices that disregard the environmental cost of a wasteful food system.[2] Food that requires less transportation to bring it to market is better for the Earth—and better for you.

Modern industrial agriculture, the source of much of our convenient and low-cost food supply, is also polluting and unsustainable and wreaks disastrous environmental impacts. In addition to toxic inputs and byproducts, large-scale agricultural practices often diminish the land's ability to support wildlife and even future crops, and disrupt the ecological properties that foster biological diversity.[3]

Factory farming has become a major public health issue. It is a leading cause of the development of antibiotic-resistant strains of bacteria.[4] Furthermore, animal waste, pesticides, and fertilizers contain nitrates, which can contaminate surface and ground

The Six Principles of Food Sovereignty[a]

1. **Food for People:** All individuals, peoples, and communities have the right to sufficient, healthy, and culturally appropriate food. Food is not just another commodity for international agribusiness.

2. **Food Providers:** All are valued who cultivate, grow, harvest, and process food. Policies, actions, and programs that undervalue them, threaten their livelihoods, and eliminate them are rejected.

3. **Localized Food Systems:** Food providers and consumers resist governance structures, agreements, and practices that depend on and promote unsustainable and inequitable international trade and give power to remote and unaccountable corporations.

4. **Local Decisions:** Food sovereignty seeks territory, land, grazing, water, seeds, livestock, and fish populations for local food providers. These resources ought to be used and shared in socially and environmentally sustainable ways that conserve diversity. Privatization of natural resources through laws, commercial contracts, and intellectual property rights regimes is rejected.

5. **Knowledge and Skill-Building:** Build on the skills and local knowledge of food providers and their local organizations, and reject technologies that undermine, threaten, or contaminate them.

6. **Working with Nature:** Use the contributions of nature in diverse production and harvesting methods that maximize the contribution of ecosystems and improve resilience and adaptation, especially in the face of climate change. Rejected are methods that harm beneficial ecosystem functions and/or depend on energy-intensive monocultures and livestock factories, destructive fishing practices, and other damaging production methods.

Source: 2007 Forum for Food Sovereignty in Sélingué, Mali signed by delegates from more than 80 countries

---

waters and have been linked to serious ailments and birth defects. Pollutants that can contaminate nearby water sources as the result of farming include sediments, pathogens, metals, and salts. These can adversely affect other ecosystems as well as the water we consume.[5]

Organic food reduces exposure to toxins like pesticide residues for those who consume them as well as for farm workers, their families and communities.

Thinking It Through

Human culture has historically revolved around agriculture. Even non-industrial agricultural societies have striven to manage and conserve food plants. And until relatively recently, organic agriculture had been the only form of agriculture practiced throughout the world.

It's only in the last century that technology began to override traditional methods of food production in favor of petroleum-based fertilizers and genetic modification. It is becoming ever more crucial for science to re-incorporate traditional knowledge and put farming back in balance with the Earth if we want to solve the distressing problems that result from large-scale commercial agriculture.

It's time to fundamentally reevaluate our food systems in light of energy efficiency, human health concerns, and environmental impact. In addition to the adverse impacts noted above, agribusiness weakens communities and family farms, creates mountains of solid waste, and funnels wealth into the hands of fewer and fewer people and corporations.

The overhaul of agriculture and food policy at the federal level could complement health care reform by encouraging better nutrition to prevent disease—which would also combat climate change. You can help by encouraging your elected officials to act.

Instead of massive subsidies—such as the $17 billion that the government spent in 2011 for growing crops used to make unhealthy food additives (corn syrup, high fructose corn syrup, corn starch and soy oils[6])—farm policy and federal dollars should encourage sustainable practices such as planting diverse crops, rewarding conservation efforts, and promoting local food networks.

GMOs

Of all the hot-button topics surrounding food, perhaps none strikes as close to the heart of so many controversial issues as genetically modified organisms, or GMOs. These crops are both hailed as being the key to feeding an increasingly hungry planet and derided as the ultimate example of industrial agriculture's disregard for human health and nature herself.

By definition, GMOs have had their DNA altered through genetic engineering, by moving genes from one species to another. In the US, GMOs—the bulk of which are patented by six corporations including DuPont, Dow, and Monsanto—surprisingly are present in up to 80 percent of processed food, either through crops such as corn, soy, and sugar beets, or through food additives like high fructose corn syrup, xanthan gum, cottonseed oil and hydrolyzed vegetable protein.

Almost all GMOs are created to be pesticide resistant or to produce insecticides themselves. Other modified crops are designed to withstand drought or contain additional nutrients. Supporters contend that GMOs pose no demonstrated threat to human health (so far) and that they are essential to meeting the food needs of our rapidly expanding population and creating crops that can withstand climate change. But GMO opponents say claims about the crops' promise are overblown. For example, a 2009 report from the Union of Concerned Scientists (ucusa. org) found that genetic engineering did not significantly increase US corn and soybean yields.

Even researchers who are critical of GMOs' commercial applications say that some claims about harm to health and environmental effects have been misrepresented. But that doesn't deny the host of potential risks associated with changing organisms' genetic makeup. These include potential new allergens and toxins, resistance to antibiotics, poisoning of wildlife, evolution of pesticide-resistant "super-weeds," and contamination of non-GMO plants by modified crops. Because of GMOs' herbicide tolerance, the use of toxic insecticides has grown 15 times since such crops were introduced, according to The Non-GMO Project (nongmoproject.org), a nonprofit that provides independent verification of products that use best practices to avoid GMOs.

GMO seeds also pose potential problems for small farms, since these seeds are subject to corporate patents and must be purchased from the manufacturer. In 2013, the US Supreme Court ruled in favor of Monsanto in a case against an Indiana farmer. The farmer, instead of buying new soybean seed from Monsanto, planted seeds derived from the previous year's soybean crop. The court affirmed Monsanto's claim that farmers must buy the company's patented seeds instead of planting seeds taken from a previous year's crop.[7]

MonsantoBoycott.com

The cost of seeds can pose financial burdens for farmers in poorer countries. This issue prompted demonstrations in Africa in the fall of 2013.[8]

The bottom line: There are so many unknowns associated with GMOs that more rigorous and independent research and approval processes, free from the influence of the GMOs' patent-holding companies, are needed. More than 60 countries, including Australia, Japan, and the European Union, have banned or restricted GMOs or mandate GMO labeling. Meanwhile, in the US at least 28 states have introduced legislation to require labeling of GMOs.

WHAT YOU CAN DO

■ Go Organic

Would you choose to dump poison on your food? Would you purposely spray farm workers with toxic chemicals? Would you intentionally eradicate butterflies or kill millions of birds annually?

When we choose to buy conventionally grown food, we indirectly do all these things. Some say that organic foods are too pricey. But the "cost" of conventionally farmed foods is clearly enormous. One option for a restricted budget is to buy organic part of the time, and reach for natural, unprocessed food for the rest. Other options include eating lower on the food chain, buying locally grown products, and joining an organic foods cooperative. Slow Food USA (slowfoodusa.org), Organic Consumers (organicconsumers.org), and The Coop Directory Service (coopdirectory.org) all offer a wide range of information on organic food.

■ Buy Local

Local food aligns us with the seasons, a sense of place, and our own commitment to reducing greenhouse gas emissions from food transport.

Farmers' markets are a great source for locally grown food, as are Community Supported Agriculture farms (CSAs). Typically, CSA members pledge in advance to contribute to the anticipated costs of the farm operation. In return, they receive shares in the farm's harvests, often on a weekly basis. To find your local farmers market, CSA, or food co-op, go to the Local Harvest website: localharvest.org.

■ Seek Out Fair Trade Products

"Fair Trade" products—notably spices, coffee, chocolate, bananas, and imported art and craft items— are those whose producers meet specific standards for treating their communities and the environment well, and for paying their suppliers a fair wage. For more information, check out the Fair Trade Federation, fairtradefederation.org, or Global Exchange, globalexchange.org.

■ Eat Less Meat

How do our food choices affect the climate? Animal agriculture is directly or indirectly responsible for many of the world's most serious environmental problems, including global warming, deforestation, air and water pollution, and species extinction.

A 2013 report from the Worldwatch Institute found that agriculture is the third-largest source of greenhouse gas emissions, after the burning of fossil fuels and transportation. Of these agricultural emissions, the largest source of greenhouse gases is the methane produced when livestock digest and excrete organic materials.[9]

Meanwhile, over half of all (increasingly scarce) fresh water consumed in the US goes to water livestock and to irrigate land where livestock feed is grown.[10]

Switching just two meals a week to a plant-based diet has a greater ecological benefit than buying all locally sourced food, according to a study in the journal Environmental Science and Technology.[11] And there's nothing to lose, nutritionally, from a vegetarian diet. The American Dietetic Association, the leading nutrition authority in the United States, affirms that "appropriately planned vegetarian diets are healthful, nutritionally adequate, and provide health benefits in the prevention and treatment of certain diseases."[12]

The farming of animals also causes them great suffering, since factory farms raise livestock in unnatural and often cruel ways to maximize their "output," with no regard to their quality of life.

In short, reducing or eliminating the consumption of meat is one of the most effective ways to heal the environment, help animals, and live a longer, healthier life.[13]

Attingham Park, Shropshire, UK; volunteers in the Walled Garden

For now, you can take steps to keep GMOs off your table if you choose. Certified organic products are prohibited from containing GMOs. Even more rigorous is the Non-GMO Project Verified label, which tells you that a product has undergone a focused third-party verification process. While potential contamination by GMO crops means that no product can absolutely claim to be GMO free, these certifications—and, of course, tapping into local farms and your own garden—are your best bet for knowing what's in—and not in—your food. For more information and a list of verified products, check out NonGMOProject.org.

DEFINING SUSTAINABLE AGRICULTURE

What is Organic Farming?

Organic agriculture is simply farming without synthetic chemicals. A pioneer of the organic movement, Lady Eve Balfour, provided a useful description: The criteria for sustainable agriculture can be summed up in one word—permanence—which means adopting techniques that:

- maintain soil fertility indefinitely

- utilize, as far as possible, only renewable resources

- do not grossly pollute the environment

- foster biological activity within the soil and throughout the cycles of all the involved food chains.[14]

Living sustainably means, in writer and environmental activist Derrick Jensen's elegantly simple definition, that whatever we do, we can do indefinitely.

Permaculture

Permaculture is a set of principles and practices for mimicking the wisdom of nature's systems and turning our homes, landscapes, and communities into productive, resilient ecosystems. This not only rebuilds self-reliance, but also addresses environmental problems—the climate crisis, topsoil loss, drought, toxic agricultural runoff (from over-fertilization), and the collapse of many wildlife populations.

Permaculture integrates ecological design, ecological engineering, and environmental design (*permaculture.org*). It's sourced in a planetary consciousness while it focuses on the areas over which we have the most direct control: our homes, gardens, relationships and selves.

We need systemic change, but systemic change starts at home. In reclaiming our power to meet more of our own needs, we can reconnect and renew ancient cycles of sowing, harvesting, and processing food together. Permaculture activities can also include composting, plant grafting, collecting rain, and using greywater (waste water from sinks, tubs, and showers that can be re-directed on site for purposes like toilet flushing and landscape irrigation).

When we meet our local needs with our hearts, hands, and neighbors, we strengthen the power of our communities. This enhances an emerging culture of stewardship and resilience—a regenerative culture that treasures every drop of water and scrap of carbon by enabling communities to catch, store, and sustain more life nutrients than they use.

Where to start is where you are, with who and what inspires you. Observe, interact, and ask: "How do we grow a regenerative culture, a permanent culture of care?" Adopting the permaculture model empowers our response to environmental and social crises. It is centered in three ethical principles: earth care, people care, and setting limits to consumption to ensure fairness.

Grounded in self-sufficiency, permaculture is being applied from American urban yards to the Jordanian desert.

Rethinking Landscaping

If you want to make a difference, why not start in your own backyard? Simple changes to your yard and garden can save water while providing food for your family and a home for animals.

Many of the grasses used in our lawns are non-native. They can succumb easily to pests and disease and require wasteful amounts of precious water to maintain. In water-stressed areas like California and the US Southwest, artificially green lawns have become an unaffordable luxury on practically every level.

Many locales are already formulating restrictions, new pricing structures, and laws to limit water use for what has come to be seen as irresponsible landscaping. The new "green" for landscaping means returning to a native ecosystem.

You can conserve water, labor, fertilizers, and pesticides by choosing "native plants"—those that naturally grow in your area with little to no infusion of water other than rain. Drought-tolerant plants are ideal. Xeriscaping refers to the use of plants that require very little water.

Specific water-friendly plant choices include succulents, groundcover for natural weed control and stability, and edibles such as strawberries, blueberries, fruit trees, vegetables, and herbs.

You could even design your yard to lower your bills. Deciduous fruit trees and vines, wisely placed, can reduce the heat in the summer and also can make up for heat loss in winter. They can thus decrease expenses and make your home more comfortable.

To make the most of your space, try dropping the one-crop-per-spot paradigm and grow a food forest of purposeful plants from canopy to root zone, using trees, shrubs, herbs, roots, and vines.

If you don't have a traditional yard or garden,

IS SEAFOOD A WISE CHOICE?

Oceans are becoming dangerously depleted by commercial fishing operations, which, in addition to their intended catch, haul in and throw away enormous amounts of sea life as "bycatch."

Fish farming operations are often toxic and unsustainable. For example, the native Pacific Northwest salmon's endangered species status is further impacted by the parasites resulting from commercial salmon farming.

Helene York, foundation director for the Bon Appétit Management Company, a food service group that provides 80 million meals a year based on a low-carbon diet, recommends clams, muscles, and oysters, which require "practically zero" energy to farm.

For help making smart and sustainable seafood choices, check out the Monterey Bay Aquarium's Seafood Watch Program at seafoodwatch.org.

container gardening can reclaim unused space, as can green walls and vertical gardens—walls either partially or fully covered in vegetation. Mostly used to control temperature in urban areas, green walls are also used to purify greywater, improve air quality, and reduce noise.

Another advantage to planting natives is reversing the trend toward disrupting natural habitats, with the resultant decline in wildlife. Birds, butterflies, and other creatures will thank you. The National Wildlife Federation offers information on backyard habitat gardening.[15]

There is no single right way. Explore, experiment, and just keep on planting. Together we are growing community self-reliance rooted in ecological resilience. We are nourishing our land as we transform waste into fertility. It's the smell, touch, and taste of the world being born, and it's right outside your door.

As Fritjof Capra, a founder of the Berkeley-based Center for Ecoliteracy *(ecoliteracy.org)* writes, nature sustains the web of life by creating and nurturing communities. Life is about relationships, and ecosystems are communities.

Tap into greywater

A good portion of household wastewater, called greywater, can be reused in landscaping. Recycling greywater helps to conserve water, lower water bills, reduce the load on sewer systems, and maintain eco-wise landscapes. Greywater may contain detergents with nitrogen or phosphorus, which are plant nutrients, resulting in more vigorous vegetation. Energy is also saved, thereby lowering your carbon footprint.

However, you'll want to research ways to minimize sodium and chloride, which can be harmful to some sensitive species. Greywater systems are site-specific. Start your online research with Greywater Action's website (greywateraction.org).

HOME SWEET NONTOXIC HOME
Creating a Haven in a Hazardous World

You may have an uneasy suspicion about how toxic and polluted our environment is—it's probably worse than you thought.

More than 85,000 chemicals are registered for commercial use in the US. Each year, the US produces or imports more than 3,000 more chemicals at quantities of over one million pounds each. Surprisingly little information is required for approval of a new chemical by the US Environmental Protection Agency.[16] Most chemicals approved for commercial use in the US have never been fully tested for their effects on human and animal health, much less test-

ed in the combinations that occur in the real world, which can increase their potency and toxicity. [17]

Newborns today begin life with almost 300 chemicals in their bodies. Lifetime health consequences of all this exposure can include cancer, allergies, asthma, skin disorders, and hormonal disruption. One in three women and one in two men will contract cancer in their lifetimes.[18,19]

European Union: The World's Environmental Leader

US chemical lobbyists often claim that removing hazardous chemicals from the products we buy would have dire economic consequences. Yet many European goods carry far fewer hazardous chemicals than their American counterparts.

In *Exposed: The Toxic Chemistry of Everyday Products and What's at Stake for American Power*, award-winning investigative journalist Mark Schapiro reveals how US companies continue to manufacture toxic-laden products in the US—even though they have reconfigured those same products *without* toxic chemicals for the European market.

The European Union has adopted the "precautionary principle" to safeguard its more than 500 million citizens. This principle *assumes* possible health and/or environmental risks instead of waiting for them to arise. Companies must be able to demonstrate that a chemical product is safe before it goes

Many US cosmetics companies still use toxic ingredients and hormone disrupters

on the market. A scientific committee regularly convenes to assess substances, posting on the EU website a growing inventory of several hundred potentially harmful ingredients.[20]

Isn't it time for the precautionary principle to be the standard in the US, as it is in Europe? It's up to us, the citizens, to insist on it.

The cosmetics industry is a flagrant example of a double standard. The EU Cosmetics Directive banned the use of chemicals determined to be carcinogens, mutagens, or reproductive toxins.[21] While many US companies redesigned their products for Europe, until recently they continued to fight potential US regulations.

Some manufacturers did eventually concede to make all cosmetics to EU standards, and that's not the only good news. In the past few years, many large cosmetic and personal care manufacturers and retailers have begun to phase out previously widely used hazardous ingredients, particularly in baby and childcare products.

But a great many US firms still use toxic ingredients and hormone disrupters. The Campaign for Safe Cosmetics *(safecosmetics.org)* encourages companies to sign their Compact for Safe Cosmetics, a pledge to replace hazardous chemicals with safe alternatives within three years of signing.

Flame Retardants: From Furniture to Children's Pajamas

The disruptive and unhealthy chemicals that regularly contact our skin include numerous chemical flame retardants. Used in upholstery, clothing, carpet backing, electronic appliance and computer plastics, and more, these chemicals' potential health hazards—particularly to children—include hormone disruption, learning and behavior deficits (including hyperactivity), birth defects and cancer.[22]

While some hazardous flame retardants have been phased out, others have taken their place. Since 2012 there has been a growing consumer movement to limit their use. Numerous US state regulations now restrict their use.[23]

Flammability standards for clothing and furniture have complicated the picture. For commonly

Everyday Green Practices

While many organizations and websites offer suggestions, the Environmental Working Group (*ewg.org*) has particularly succinct Guide Sheets for citizens to make informed choices. Some highlights from EWG:

Choose better body-care products. "Gentle" or "natural" or "nontoxic" have no legal definition or guidelines. Read the ingredients. Avoid triclosan, BHA, fragrance, formaldehyde-releasers, phthalates and oxybenzone. Check out the EWG Skin-Deep Cosmetics Database for info on many products and grades on companies. (*ewg.org/skindeep*)

Choose plastics carefully. Avoid clear, hard plastic bottles marked with a "7" or "PC" and toys marked with a "3" or "PVC." Give your baby a frozen washcloth instead of vinyl teethers.

Filter your tap water. There are many home water filters on the market, including reverse osmosis systems, alkaline and ionizing water filters, or a simple carbon filter pitcher. Don't drink bottled water—save your money and save the planet.

Cook with cast iron or stainless steel only. Nonstick can emit toxic fumes and leach its chemical coatings into your food.

Use a HEPA-filter vacuum. Kids spend a lot of time on the floor, and household dust can contain contaminants like lead and fire retardants.

Have your home checked for lead paint if your home was built and painted before 1978. If needed, local health authorities can help you find the right resources for testing and removal. If your home had outdoor lead paint, also have your soil checked, especially if you have children playing around the house or if you plan to start a garden.

Get your iodine. Use iodized salt. Iodine buffers against chemicals like perchlorate, which can disrupt your thyroid system and affect brain development.

Do not heat plastics in a microwave. This causes chemicals to leach from plastic into food and beverages.

Avoid anything plastic for infants and young children.

House cleaning: The Ecology Center in Berkeley, CA (ecologycenter.org) offers a Fact Sheet of Alternative Cleaning Recipes. You'll notice that nontoxic alternatives are less expensive.

used plastics, foams and synthetics to meet these standards, manufacturers have had to use chemical flame retardants. California recently changes its furniture flammability standard to enable manufacturers to meet its requirements without these chemicals.

But if you don't want toxic flame retardants touching your children's skin, you'll want to choose products made from materials such as cotton and wool, which don't include chemical fire retardants.

As for upholstery, avoid contact with chemical flame retardants by purchasing furniture that has no foam and that's covered by naturally fire-resistant fabrics such as cotton or wool. The *National Green Pages (greenpages.org)*, published yearly by Green America, lists companies offering nearly every product and service needed for a healthy life.

THE ECOLOGICAL FOOTPRINT

Squaring humanity's demand with nature's supply

Just like any company, nature has a budget. It can only produce so many resources and absorb so much waste every year. The problem is, we throw away more than nature can absorb, and we extract more than it can regenerate.

Using a resource accounting tool called the Ecological Footprint, we can measure the land area it takes to produce the resources a population consumes and to absorb its waste. The Ecological Footprint compares human demand against biocapacity—what nature can supply—just as financial accounting tracks expenditures against income. And the current ledgers are sobering.

According to the Global Footprint Network, a research institution that calculates the Ecological Footprint for 150 nations, globally we demand the resources it would take one and a half Earths to renewably produce. Put another way, it takes about 18 months for nature to produce the biological resources humanity demands in one year.[24] So we're cutting down trees faster than they re-grow, and catching fish faster than they repopulate.

This means we're not living off the Earth's "interest." We're irresponsibly draining "principal." And as any household budgeter knows, that's not a sustainable situation. It can be done for a short while, but we are in the process of literally bankrupting nature, on which all life depends.

The Ecology Center/Scott Sporleder

Ecology Center Eco House is a demonstration home and garden located in a Berkeley residential neighborhood

While globally we are demanding the resources of 1.5 planets, some countries demand much more and some much less. In the US, the average person's Ecological Footprint, meaning how much ecologically productive land and water one person requires the equivalent of 17 football fields.[25]

At the other end of the spectrum are countries like Haiti, Bangladesh, and Malawi, with Ecological Footprints of less than 1.3 global acres per capita—in most cases, too small to provide for the basic needs for food, housing, and sanitation.

Although high-income nations are clustered at the high end of the Footprint scale, nations with similar standards of living—as measured by longevity, income, literacy rate, child mortality, etc.—can have very different levels of resource consumption. The average European, for example, has a footprint half that of the average American.

Why is this the case? The answer lies partially in the way our societies are structured. Consider Italy, which has a per capita footprint half that of the US. Most people live in compact cities, where they can walk to work, school, and shopping, or use buses and trains. Public transportation is easily accessible—often more convenient and inexpensive than driving.

People get much of their food from local markets and food producers, and eat less packaged and frozen food. Also, since homes in cities often share walls and thus have less exposure to the outside, they consume less energy for cooling and heating.

Some of the US's gigantic Ecological Footprint is due to individual consumption choices we make, which are completely within our control, such as:

- driving instead of riding the bus

- living in a detached home rather than a multi-unit dwelling

- buying more stuff than we need

- eating meat, which takes many times more energy to produce than vegetables and fruits, and

- buying food that's been shipped from thousands of miles away.

Much of our Ecological Footprint, however, is the result of infrastructure decisions made by business leaders and policymakers, in some cases decades ago: decisions such as investing in highways rather than public transportation, and creating suburban sprawl rather than concentrated, urban development.

For that reason, one of the most important individual actions we can take is to urge our business and government leaders to make decisions that will help balance our "budget" with nature, not further aggravate the debt.

You can determine your personal ecological footprint and what actions you can take to reduce it at footprintnetwork.org/calculator.

The Way Forward

While the realities we face are indeed troubling, we have key opportunities to reverse current trends. On a community level, creating resource-efficient cities and infrastructure, fostering best-practice green technology and innovation, and making our Ecological Footprint central to decision-making at all levels of leadership can begin to turn the tide.

Human ingenuity has transformed the way we use nature. We must now put that talent toward another transformation: creating a society that truly understands that nature's fate is our fate. Prosperity and opportunity can only be a function of what the planet can sustainably provide.

There is no economy on a dead planet; human civilization is dependent on a healthy planet.

THE CONSUMPTION CONUNDRUM

Let's tell this one with an economy of words: We in the West buy and throw away way too many things. Our appetite for everything from sports cars to swordfish is driving climate change, mass species extinction, and an inequitable economic system. The widespread American assumptions that consumption is a core value, progress equals growth, and more is better, are all myopic and fallacious, not to mention a root cause of environmental degradation.

Collapse or change of this economic model is a

100% certainty. We cannot go on living this way on a finite planet. Rapidly growing energy and resource consumption is consistently breaking records on this planet. The global ecosystem upon which life depends is being stretched to its limits.

While environmental and social, and other challenges are being addressed in innovative ways, changes in consumption patterns lag behind. High-consumption nations must embrace simplicity as opposed to extravagance, community as opposed to competition, the ethereal as opposed the material. By embracing these truths, the real work can begin.

Consumer Spending Can't Save Us

Commercialism and retail therapy won't save our economy, let alone our ecosystem. The standard remedy of getting consumers or the government to spend more can no longer work because the planet is telling us, loud and clear, than it can't cope with business as usual. Whatever government and consumers spend needs to reflect that reality.

What's the Story with "The Story of Stuff"?

An Internet sensation, "The Story of Stuff" is a 20-minute animated video—seen by over 40 million people around the world—that takes a revealing look at our production and consumption patterns and the dysfunctional global economy we are enmeshed in.

The corresponding website has more great videos, resources and learning tools, and lists NGOs working on the intertwined issues of extraction, production, distribution, consumption, and disposal. Check it out at storyofstuff.org.

It's estimated that 5000 tons of trash accumulate each year in the Grijalva River, Chiapa de Corzo, Chiapas, Mexico

Economist Juliet Schor advises more sharing: job sharing, property and income redistributing, and sharing of access and know-how. "This time the economic pain needs to be assuaged by deeper structural changes that reintroduce fairness into our system. That's not just moral, it's also good economic sense."[26]

Where should our dollars be spent? On purchases that enhance and regenerate the Earth and its inhabitants, on businesses that are truly sustainable, and in support of nonprofit groups doing important life-sustaining work. In short, private and public funds alike must go to the green economy.

But what is also necessary is moving toward "voluntary simplicity"—consuming only what is needed—and restructuring the economy for shared distribution of the fruits of our society.

A child born today in an industrialized country will pollute and consume more over his or her lifetime than 30 to 50 children born in developing countries.[27] With judicious use of our purchasing power, we can shift these institutions and habits toward sustainability. The Center for a New American Dream (newdream.org) offers a variety of tools for re-conceiving our cultural patterns of consumption and choosing sustainable practices.

A Culture of Simplicity

The voluntary simplicity movement envisions a new definition of progress: living in balance with self, other and bioregion, deriving a sense of fulfillment from relationships and nature. The focus is on developing talents, creativity and community rather than status and materialism, placing high value on social issues and creation of a better society.

As we move beyond a cultural obsession that equates happiness with material possessions, we become more and more unwilling to sacrifice quality of life, equality of opportunity, and even our health for meaning and status based on accumulation.

WHAT YOU CAN DO

Practical ways to lower your consumption and spare the Earth

- Power down: This includes shutting down your computer when not in use, at work and at home, as well as unplugging the TV and Internet and plugging in to community instead.

- Consume less, waste less. Especially:

  • Don't drink bottled water.

  • Bring a reusable bag wherever you go, not just to the grocery store.

  • Eat less processed food. Processed foods require high inputs of energy in processing, packaging, and transportation, not to mention the packaging waste involved.

- Reduce or eliminate your meat consumption. Again, it takes ten times as much energy to produce a pound of meat than a pound of vegetables.

- Make your own cleaning products (see "Home Sweet Nontoxic Home" above).

- Eat more meals at home. Grow some of your own food. That's about as local as it gets.

- Consider the embodied energy of every article you use and consume, meaning its cost to be made, stored, and carried to where you are. Get clarity on your wants versus your needs.

- Park your car and walk.

- Change your lightbulbs … and then, change your paradigm.

- Recycle, reuse, and repurpose your items.

- Overall, buy green, buy fair, buy local, buy used, and most importantly, buy less — voluntary simplicity is the most effective way to reduce consumption and spare the Earth.

SPIRITUALITY

A Personal Compass for Today's World

"In a real sense, all life is inter-related. All persons are caught in an inescapable network of mutuality, tied in a single garment of destiny. Whatever affects one directly, affects all indirectly. I can never be what I ought to be until you are what you ought to be, and you can never be what you ought to be until I am what I ought to be."

—Martin Luther King, Jr.

The times we live in are more challenging than most of us are willing to face. Increasingly, the scope, scale, and urgency of our global crises are becoming devastatingly clear. Financial instability, terrorism, climate change, the destruction of our natural world, widespread social injustice. In the face of this dizzying array of crises, we naturally feel anxiety or despair, lost in the apparent chaos, handcuffed by our perceived limits and our vulnerabilities.

It's when things get most difficult that we ask the existential questions: Is life an ordered system where good triumphs and justice prevails, or a mere roll of the dice? Is it survival of the fittest, or does karma reward the good? When the answers don't appear, a crisis of spirit can arise.

To rise above despair and engage positively in today's world, we have to find a way to feel connected and worthy, and to know against all odds that we are not powerless. One reliable path to hope and energy is to make active contributions—to our family, community, or civil society. To dedicate our lives to something larger than ourselves.

When Going Within Means Going Away: In the face of profound and frightening change, people often contract spiritually and emotionally to stave off fear, anger, loneliness, and grief. But this means losing our connection with other humans and with the natural world. We lose our capacity for empowering visions and creative solutions.

We see this played out in addictions, suicides, depression, and seemingly weekly incidents of mass shootings. All result from disconnection and isolation—the devastating feeling that we do not belong to each other.

Make a Space for What's Good in the World: Some of us act out of anger and fear because we have not

Estancia Ranquilco, Patagonia, Argentina

Eddie Thornton

The best way to tell whether we are moving in the direction of greater wellbeing is by listening to our inner messages of comfort or distress. Our highest evolutionary path is the one that generates the least resistance and the most joy."

—*David Simon, M.D.*

embraced our grief, which sits just below the surface. In trying to deny our pain, we actually increase its impact on us.

Yet, underneath it all, intuitively each of us knows that life will survive. There is every reason not to go unconscious and disconnect, to spiritually cop out. Within the nature of evolution is the inherent drive toward the good, the true, and the beautiful. If we look at Earth's history, this trajectory is clear in infinite ways.

To survive, to thrive, we must consciously choose to evolve by changing our beliefs and behaviors. Yes, the world is experiencing unprecedented crises, but crisis births transformation. And when we transform our own individual "crisis" response, spiritual fulfillment dawns. As Gandhi said, "If we want to remake the world, we must remake ourselves."

So what does is mean to be spiritually fulfilled mean? While spiritual teachers may have different answers, all agree that compassion and true inner peace are crucial to both personal and planetary transformation.

We are solitary creatures, but what we yearn for most, and in fact what we're biologically programmed for, is connection—to ourselves, to each other, and to our world. To deeply experience these layers of connection, we must commit to our own inner development. We must embrace our negative feelings, whatever form they take, and transform that energy into strength, passion, compassion, and commitment—into an embodied conviction that every life matters, and *my* life, which inevitably belongs to the whole, can make a difference.

There is no need for temples, no need for complicated philosophies. My brain and my heart are my temples; my philosophy is kindness.

—*The Dalai Lama*

Steps to Change from the Inside

Reflecting on powerful questions provides access to wisdom. Here are a few to ask ourselves:

- What am I spending? What am I being spent for?

- What commands and receives my best time, my best energy?

- What and whom am I committed to in life?

- With whom do I share my most sacred and personal hopes for my life, and for the lives of those I love?

- What are my most compelling life goals and purposes?

(adapted from Stages of Life *by James Fowler)*[b]

Living into these questions will help us move in the direction of our own transformation and spiritual fulfillment. Like all things, though, we can't **fully** do this alone. Like-minded community is essential to support us on the way. And spiritual fulfillment is a two-fold path—the inner path of our own personal journey, and the outer path of compassion and caring for others.

EXPLORE & ENGAGE

Individual efforts may feel trivial when facing the enormity of the challenges. Yet both modern science and ancient wisdom tell us that the flap of a butterfly's wings or an act of kindness can shift the universe. One action—and we can never know in advance *which* action—could be the tipping point that makes all the difference in the course of history. Each of us has the potential to make a powerful contribution...and if enough of us are committed to being the change, we can and will create a just, thriving, and sustainable world.

> Never be afraid to try something new. Remember, amateurs built the ark; professionals built the Titanic.
>
> — *Unknown*

1. Our food choices have broad implications.

ⓠ   How do food choices impact our health, land, energy, and social justice? Contrast the impact of a typical US grocery cart with three reusable bags full of local, organic, vegetarian food.

ⓠ   Discuss the many factors that influence your current eating habits. How might your food choices be influenced by advertising?

- This short video shows how fast-food commercials are prepared behind-the-scenes by a "food make-up artist" to generate consumer desire. ◉

➡   Make a list of changes that would better support your health and your values.

➡   Choose one of the Six Principles of Food Sovereignty (p.130) and take one action to support that principle.

📖   As you shop for food (and eat out) this week, identify as best you can where your food comes from. How does the food you buy rely on long-distance transport?

➡   With your family or people you eat with, discuss how to eat more local produce.

➡   Read one of the following books and report back:

- *Animal, Vegetable, Mineral* (Barbara Kingsolver)
- *The Omnivore's Dilemma* (Michael Pollan)
- *Diet for a Small Planet* (Francis Moore Lappe)
- *Diet for a New America* (John Robbins)

2. Eating organic food supports your health and that of the planet.

ⓠ What is "even better than organic food"? (p.132)

📖 Discover what sources of local or organic food are available in your area. Include farmers' markets and CSAs.

- This TED talk video features an 11-year old promoting localized organic foods. 👁

➡ Go to a farmers' market and spend some time talking to the vendors about their crops. Make a point of speaking with farmers whose crops are "in transition."

➡ Start a "foodie" group to support local organic food. Consider having potlucks, gardening parties, Slow Food events, and book readings.

3. The products you use in your home have an impact on your health and the planet.

ⓠ Read the following facts out loud:

- Newborns begin life with almost 300 chemicals in their bodies.

- 40% of US rivers, lakes, and coastal waters are so contaminated that they are unfit for humans to fish in, swim in, or drink.

- More than 90% of the 85,000 chemicals in use today have never been tested for their effects on human health.

- US companies have reconfigured products without toxic chemicals for the European market while continuing to manufacture the same products in the US *with* toxins.

- PBDE, used in flame-retardants, is found in alarming concentrations in human blood and breast milk—affecting learning, memory, and regulation of hormones.

➡ This video thoroughly discusses the chemical contamination in the cycle of food products, from start to finish. 👁

💡 Sit in silence for a minute. Become aware of emotions that arise as you hear these facts. Next, notice subtle responses you may have in your body. Finally, pay attention to thoughts that show up.

ⓠ Share your feelings and reactions with others.

➡ Conduct an informal survey of 2-4 people outside your group to determine which of these facts they were aware of. Discuss your results with your group.

4. North America's Ecological Footprint (EF) is significantly larger than that of Western Europe.

Big Stock Photo

ADD TO CART

📖   Determine your own EF. ◉

⑦   Which of the factors that contribute to your personal EF are under your control? What societal factors might you be able to influence?

⑦   Discuss some of the factors that account for the difference between the continents.

⑦   Asia is now approaching the *1 earth* level. (p.69) What are the implications of that?

➡   Commit to 2-3 actions in the next month that will improve your personal, community, state, and/or national EF.

5. Consumerism is becoming a powerful force in many countries.

⑦   Discuss the short-sightedness of linking economic recovery to increasing consumer spending.

➡   Watch *What a Way to Go: Life at the End of Empire* ◉ as a group (two hours). It offers powerful insights on the "American Dream." Share your reactions.

📖   Read about "The 100 Things Challenge" and calculate how many things you have. ◉

⑦   Brainstorm ways to live more simply and reduce your consumption.

⑦   How does "voluntary simplicity" differ from simply doing without things? (p.141) How is living sustainably different from living a deprived life?

➡   Write a blog or otherwise share the changes you are making. Acknowledge emotions and challenges as well as successes.

6. The demand for bottled water has largely been created by the bottled water industry.

Creative Commons/Grand Canyon National Parks Service

SPRING WATER
FILL YOUR BOTTLES HERE FOR FREE

Refill your bottles here and enjoy pure, clean Grand Canyon spring water.

Kids, did you know...

📖   Watch *The Story of Bottled Water.* ◉ Discuss how an entire industry was developed based on creating an artificial demand. How is the process similar to creating a market for processed food?

♥   Group Exercise: Positions on Bottled Water

•   Divide the group into two teams.

•   One team assumes the role of people in developed countries who regularly buy bottled water. The other team represents people living 20 years ago who relied on tap water—filtered or otherwise.

- Each team's objective is to convince the other to adopt its practices.

- Each team will show appreciation and respect for the members of the other team.

- Debrief by discussing your experience and what you learned.

➡ Make a commitment as to what you will do to reduce the use of bottled water in both your life and in that of your sphere of influence—your friends and co-workers; restaurants, stores, etc.

Nick Amgler

A bird doesn't sing because it has an answer, it sings because it has a song.

—Maya Angelou

7. Spirituality is an aspect of our personal relationship with sustainability.

💡 Studies consistently show that there is little connection between happiness and material possessions, once one has the basics of food, clothing and shelter. Reflect on what this truth means to you. Examine how you have used stuff and entertainment to feel better or to minimize loneliness or sadness. Sit for a moment in silence and drop your focus from your head into your heart. Allow yourself to feel gratitude or joy for the happiness you do have in your life.

❓ What does spiritual fulfillment mean to you? How do you create connection in your life with your innermost self, with other people, and with the earth? Discuss.

❓ What does "The Great Turning" mean to you?

❓ Share what you consider to be your spiritual practices with the group. What is your edge—what could you do to be more consistent with your practices, or go deeper into them?

❓ How do you see sustainability as part of spirituality? Can one's commitment to living a life that promotes envirnonmental protection and social justice be considered part of one's spiritual practice?

❓ How do different major religions address sustainability and social justice?

Mama exhorted her children at every opportunity to "jump at the sun." We might not land on the sun, but at least we would get off the ground.

—Zora Neale Hurston

RESOURCE DIRECTORY

"Every few hundred years in Western history there occurs a sharp transformation. Within a few short decades, society—its world view, its basic values, its social and political structures, its arts, its key institutions—rearranges itself. And the people born then cannot even imagine a world in which their grandparents lived and into which their own parents were born. We are currently living through such a transformation."

—Peter Drucker, author, Post-Capitalist Society

Welcome to the vast array of organizations we invite you to sample for deeper research, links to videos, action programs, etc. Organized by relationship to *Sourcebook* chapters, the organizations listed represent a sample of the amazing work going on (the majority are based in the US, but with the web they can be accessed from anywhere). Give us your suggestions of other useful resources at info@swcoalition.org, and access the searchable pdf version of the *Sourcebook*, where all the links are active, at swcoalition.org.

ENVIRONMENT

USGS

350.org
Global climate change activist organization, with a mission of keeping CO_2 levels below 350 ppm, known to be the safe limit. 350.org

Acterra
Palo Alto, CA; Protecting the environment with restoration and education. acterra.org

Alliance for Climate Education (ACE)
Various US cities; Educates high school students on the science behind climate change and inspires them to take action. acespace.org

Amazon Watch
San Francisco, CA; Defends the environment, rights of indigenous peoples of the Amazon basin. amazonwatch.org

American College and University Presidents Climate Commitment
Boston, MA; Modeling ways on campuses to minimize global warming emissions and achieve climate neutrality. presidentsclimatecommitment.org

Association for the Advancement of Sustainability in Higher Education
Denver, CO; Info on campus sustainability, discussion forums, initiatives & consortiums. aashe.org

Biomethane.com
All about biomethane—links to related info, products and uses. biomethane.com

Blue Green Alliance
Broad coalition catalyzing a clean-energy revolution in the US and creating many green-collar jobs. bluegreenalliance.org

Blue Planet Project
Ontario, Canada; Works with other organizations around the world for the right to healthy water and sanitation. blueplanetproject.net

California Student Sustainability Coalition
Berkeley, CA; Connects, supports and empowers students in California to transform their schools and communities into models of sustainability. sustainabilitycoalition.org

Carnegie Institution for Science
Stanford University, CA; Research, publications and seminars about the earth's ecosystems. dge.stanford.edu

Center for Biological Diversity
Tucson, AZ; Working through science, law, media for all species on the brink of extinction. biologicaldiversity.org

Center for Health, Environment and Justice
Falls Church, VA; Supports grassroots involvement in environmental policies. chej.org

Climate Reality Project
Educates about climate change and solutions to the climate crisis; founded by Al Gore. climaterealityproject.org

College Sustainability Report Card
Interactive website providing sustainability profiles of colleges in the U.S. and Canada. greenreportcard.org

Conservation International
Arlington, VA; Supports worldwide economic and infrastructure work for solutions to protect planet resources. conservation.org

Earthday Network
Washington, DC; Works with activists worldwide for environmental awareness and policy changes. earthday.org

Earth Island Institute
Oakland, CA; A non-profit organization that supports people creating solutions to protect the planet. earthisland.org

Earth Justice
Washington, DC; Public interest nonprofit law firm dedicated to a healthy environment. earthjustice.org

Earth Policy Institute
Washington, DC; A plan for a sustainable future, along with a roadmap to get there; founded by Lester Brown. earth-policy.org

Ecology Action
Santa Cruz, CA; Education and programs on environmental awareness, change, and sustainable economy. ecoact.org

Endangered Species Coalition
Washington, DC; Educates on ways to protect endangered species and their environments via grassroots mobilization. endangered.org

EnviroLink Network
Pittsburgh, PA; Provides environmental information and news from thousands of orgs around the world. envirolink.org

Environmental and Energy Study Institute
Washington, DC; Educates policymakers and develops initiatives for an environmentally sustainable America. eesi.org

Environmental Defense Fund
Washington, DC; Preserving natural systems, focusing on the most critical environmental problems. edf.org

Environmental News Service
Oregon and Taiwan; Environmental news from around the world, from journalists and experts in relevant fields. ens-newswire.com

European Environment Agency
EU information agency for developing, implementing and evaluating environmental policy. eea.europa.eu

Food and Water Watch
Washington, DC; Advocates policies for healthy and sustainable water and food for all people of the world. foodandwaterwatch.org

Friends of the Earth
Washington, DC and the Netherlands; Focusing on clean energy, global warming, toxin protection, smarter transportation. foe.org (US), foei.org (international)

Global Footprint Network
Oakland, CA; Provides information and tools regarding the world's ecological footprint. footprintnetwork.org

Greenpeace USA
San Francisco, CA & Washington, DC; Defends the natural world and promotes peace through direct action and environmentally responsible solutions. greenpeace.org/usa

Greywater Action
Educating, empowering to build sustainable water culture, infrastructure. greywateraction.com

Harvest H2O
Santa Fe, NM; Advancing sustainable water management practices. harvesth2o.com

Health Care Without Harm
Promotes ecologically sound alternatives to healthcare practices that pollute the environment. noharm.org

ICLEI Local Governments for Sustainability
Helping over 1200 local government members to implement sustainable development. icleiusa.org

Indigenous Environmental Network
Bemidji, MN; Links grassroots and tribal governments protecting the sacredness of water. ienearth.org

International Rivers
Berkeley, CA; Works to halt destructive river projects worldwide, and to support equitable sustenance. internationalrivers.org

International Scientific Congress on Climate Change
San Francisco, CA; Providing a synthesis of existing and emerging scientific knowledge to make sound decisions. globalwarmingisreal.com

ItsGettingHotinHere.org
Worldwide; Global online youth community reporting about climate change. itsgettinghotinhere.org

Jane Goodall Institute
Vienna, VA; Programs to support worldwide understanding and preservation of the environment. janegoodall.org

Living Planet Index
Worldwide; Measures trends in the Earth's biological diversity; from World Wildlife Fund International. panda.org

Marine Mammal Center
Sausalito, CA; Appreciation of marine mammals through rescue and treatment, scientific inquiry, education and communication. tmmc.org

National Audubon Society
New York, NY; Education, political action, and grassroots efforts to promote conservation and biodiversity. audubon.org

National Geographic
Washington, DC; Multi-media producer exploring, educating and conserving the natural world around us. nationalgeographic.com

National Park Conservancy Association
Washington, DC; Safeguarding America's scenic beauty, wildlife, historical treasures. npca.org

National Wildlife Federation
Merrifield, VA; Promotes conservation, encourages social diversity, and demands policy change. nwf.org

Natural Resources Defense Council
New York, NY; Advocates for environmental protection through mobilization of grassroots efforts, policy initiatives and research. nrdc.org

Nature Conservancy
Arlington, VA; Coalitions, research, campaigns, grassroots work for conservation of the environment worldwide. nature.org

NoNukes.org
Global library, links to info about nuclear power, weapons, contamination, citizen action. nonukes.org

Northwest Seed
Seattle, WA; Sustainable energy projects and education for communities. nwseed.org

Ocean Arks International
Burlington, VT; Global leader in the field of ecological water purification. oceanarksint.org

Ocean Futures Society
Santa Barbara, CA; Educates about the significance of oceans and promotes actions to protect oceans. oceanfutures.org

Project Kaisei
San Francisco & Hong Kong; Studying the North Pacific Gyre plastic debris. projectkaisei.org

Public Interest Research Groups (PIRGs)
Hundreds of grassroots groups disseminating info on justice, environment. pirgs.org

Rainforest Action Network
San Francisco, CA; Transforming the global marketplace via education, organizing, non-violent direct action. ran.org

Rainforest Alliance
New York, NY; Protects ecosystems, people and wildlife. rainforest-alliance.org

Sarvodaya Shramadana Movement
Sri Lanka; Involved in resettlement, reconstruction and reconciliation activities in war affected areas. sarvodaya.org

Sarvodaya Shramadana Movement
Sri Lanka; Involved in resettlement, reconstruction and reconciliation activities in war affected areas. sarvodaya.org

Second Nature
Boston, MA; Supports national initiatives and programs aimed at reorienting the higher education sector toward more sustainable outcomes. secondnature.org

Species Alliance
Emeryville, CA; Educates about endangered species to stimulate public policies and inspire healthy human behavior. speciesalliance.org

Student Conservation Association
Arlington, VA; Hands-on conservation programs for youth. thesca.org

Sustainable Endowment Institute
Boston, MA; Initiatives for nonprofit support of programs promoting environmental sustainability. endowmentinstitute.org

Suzuki Foundation
Canada; Finds ways for society to live in balance with the natural world. davidsuzuki.org

Tides Center
San Francisco, CA; Provides financial aid to nonprofits around the U.S. and charitable organizations worldwide. tidescenter.org

Trust for Public Land
San Francisco, CA; Works nationwide to help agencies and communities conserve land for public use. tpl.org

United Nations Environmental Program
Nairobi, Kenya; Provides leadership and partnership strategies to aid in global environmental management. unep.org

United Nations GEO Report
Global Environment Outlook provides overview and reports on state of environment and policy. unep.org/geo/geo3

Water Resources Research Center, University of Arizona
AZ; Promotes understanding of critical state and regional water management and policy issues. ag.arizona.edu

Water Institute
Information and resources for local watershed activism. oaec.org

Water Use It Wisely
100 ways to conserve water; reports, videos, links to water conservation products. wateruseitwisely.com

Wild Gift
Hailey, ID; Education, seed money, training for worldwide projects to resolve global environmental challenges. wildgift.org

Wilderness Society
Washington, DC; Conservation of wilderness lands through policy initiatives and education. wilderness.org

Wind Energy Institute
Educate consumers about wind power and other renewable energy. windenergyinstitute.com

World Resources Institute
Washington, DC; Global environmental think tank of research, resources and active projects. wri.org

World Wildlife Fund
US & Switzerland; Working to preserve the diversity and abundance of life on Earth. worldwildlife.org

Worldwatch Institute
Washington, DC, Accessible, fact-based analysis of critical global issues in order to create an environmentally sustainable society. worldwatch.org

Zero Waste Alliance
Portland, OR; Provides resources and works with all aspects of communities, business and government toward zero waste solutions. zerowaste.org

ENERGY

American Council for an Energy-Efficient Economy
Washington, DC; Advancing energy efficiency as a means to economic prosperity. aceee.org

American Council on Renewable Energy
Washington, DC; Educational resources for technology, finance and policy relating to renew-able energy solutions for America. acore.org

Build It Green
Oakland, CA; Resources and education to promote efficient and healthy buildings in California. builditgreen.org

California Cars Initiative
Palo Alto, CA; Promoting efficient, non-polluting automotive public policy and technologies. calcars.org

Citizens Climate Lobby
US nationwide; Putting a price on carbon and creating the political will for a stable climate. citizensclimatelobby.org

EcoGeek
Technology news, tips, and blogs on energy, policy, green food, and alternatives to cars. ecogeek.org

Elevate Energy
Chicago, IL; Designs and implements energy efficiency programs, particularly for the underserved. elevateenergy.org

Energy Action Coalition
US and Canada; Youth coalition for clean energy. energyactioncoalition.org

Energy and Resources Institute
New Delhi, India; Environmental problem-solving think tank that disseminates the resulting research.

Green Building Councils
Woodbridge, ON, Canada; Helping the construction industry reduce GHG emissions from the built environment. worldgbc.org

Home Energy
Berkeley, CA; News on energy-efficient, durable, comfortable, and green homes. homeenergy.org

Intergovernmental Renewable Energy Organization (UN)
New York, NY; promotes the urgent transition to sustainable development and renewable energy sources. ireoigo.org

Creative Commons/Think Panama

National Renewable Energy Laboratory
Washington, DC; Basic information about renewable energy for consumers, homeowners and businesses. nrel.gov

Native Energy
Charlotte, VT; Offering Native Americans certified RECs and CO_2 offsets that help build new renewable energy projects. nativeenergy.com

Plug-In Scam
Yellow Springs, OH; Blog about electric vehicles, particularly to expose government misrepresentation of plug-in MPG ratings. pluginscam.org

Post Carbon Institute
Sebastopol, CA; Transition from fossil fuels: research and education. postcarbon.org

Renewable Energy Institute
McClellan, CA; Archive of research for technologies to convert waste biomass and coal to fuels. reiinternational.org

Rocky Mountain Institute
Providing R&D, funding, resources and design that drive the efficient and restorative use of resources. rmi.org

StopGlobalWarming.org
Pacific Palisades, CA; Promotes knowledge about global warming and grassroots solutions. stopglobalwarming.org

US Department of Energy Efficiency and Renewables
Washington, DC; Works with organizations and industry to promote energy efficient practices. efficiency info. eere.energy.gov

Wind Energy Institute
Educate consumers about wind power and other renewable energy. windenergyinstitute.com

World Green Building Council
Toronto, Ontario; Global organization that works with the building industry and governments for sustainable policies and methods. worldgbc.org

A JUST SOCIETY

Alternative Information and Development Centre
Capetown, South Africa; Works in Africa for economic justice and social transformation. aidc.org.za

Amnesty International
London, UK; Preventing human rights abuse and demanding justice. amnesty.org

Applied Research Center
New York, Oakland, Chicago; Racial justice research, policy, media, activism. arc.org

Asian American Justice Center
Washington, DC; Advancing human and civil rights, with a large network of community-based organizations. advancingequality.org

Asian Law Caucus
San Francisco, CA; Promoting legal and civil rights of Asians and Pacific Islanders. asianlawcaucus.org

Asian Pacific Environmental Network
Oakland, CA; Supports environmental, social and economic justice for Asian and Pacific Islander communities. apen4ej.org

Bidwell Training Center
Pittsburgh, PA; Academic training with financial aid and support particularly for low-income adults in transition. bidwell-training.org

Boggs Center to Nurture Community Leadership
Detroit, MI; Multi-cultural community activism encouraging strategies for rebuilding our cities. boggscenter.org

The Center for Media Democracy
Madison, WI; Promotes media transparency and informed debate, engaging the public in collaborative, fair and accurate reporting. prwatch.org

Centre For Child Honouring
Canada; Communicates and educates about the significance of a healthy world community to nurture children in a safe and ethical manner. childhonouring.org

Children and Nature Network
Santa Fe, NM; Mission to reconnect children with nature through resources and tools. childrenandnature.org

Color of Change
Oakland, CA; Empowering Black Americans towards positive political and social change. colorofchange.org

ColorLines
New York, NY and Oakland, CA; News research and reporting from a community, social justice and multiracial perspective. colorlines.com

Congress of Racial Equality
New York, NY; Acts to uncover discrimination and to empower minority civil rights around the world. core-online.org

Congressional Black Caucus Foundation
Policy research and education to improve socioeconomic circumstances of underserved communities.

Cultural Creatives
CA; Uniting 50 million adults in the US who share worldview, values and lifestyle.

Cultural Survival
Cambridge, MA; Promotes the rights, voices, and visions of indigenous peoples. culturalsurvival.org

Delancey Street Foundation
San Francisco, CA; Residential self-help organization for homeless, ex-convicts, substance abusers. delanceystreetfoundation.org

EarthJustice
San Francisco, CA; Works in the U.S. to give free legal assistance to preserve wildlife and healthy environments, to promote clean energy and combat global warming. earthjustice.org

Edible School Yard
Berkeley, CA; Online interactive resource for schools and other groups to bring children into a positive relationship with their food. edibleschoolyard.org

Ella Baker Center for Human Rights
Oakland, CA; Resources and advocates for empowering low-income people and people of color to work together for strong and healthy communities. ellabakercenter.org

End Poverty UN 2015 Millennium Campaign
New York, NY; International commitment promoting citizen action to end poverty by 2015. endpoverty2015.org

Environmental Justice Resource Center at Clark Atlanta University
Atlanta, GA; Assists people of color to be included in decision-making of environmental issues. ejrc.cau.edu

Environmental Working Group
Washington, DC; Protecting public health; pushing for national policy change. ewg.org

Equality Now
Works to end violence and discrimination against women and girls around the world, including female genital mutilation. equalitynow.org

Fairness and Accuracy in Reporting
Well-documented criticism of media bias, censorship. fair.org

Federation of Southern Cooperatives
Works for sustainability and land retention for poor southern U.S. family farmers, particularly African Americans. federationsoutherncoop.com

Free Press
Promoting diverse, independent media ownership, strong public media, quality journalism. freepress.net

Genesys Works
Houston, TX; Helps disadvantaged youth gain experiences and knowledge required for success in the corporate world. genesysworks.org

Global Exchange
San Francisco, CA; Campaigns around the world for social and economic justice. Co-producer of Green Festivals. globalexchange.org

Global Restoration Network
Washington, DC; Database and portal for ecological restoration studies. globalrestorationnetwork.org

Green For All
Working to build an inclusive green economy to lift people out of poverty. greenforall.org

Growing Power
Milwaukee, WI; Building equitable, ecologically sound, sustainable food systems, one community at a time. growingpower.org

Harlem Children's Zone
Harlem, NY; Replicable holistic system of education, social-service and community-building. hcz.org

Health Care Without Harm
Reston, VA; Worldwide coalition of healthcare organizations and individuals with supportive resources to provide environmentally safe healthcare. noharm.org or hcwh.org

Heifer International
Little Rock, AR; Helping to end hunger and poverty through providing livestock as source of food and self-reliance. heifer.org

Highlander Research and Education Center
New Market, TN; Supports grassroots organizing for justice and sustainability in Appalachia and the South. highlandercenter.org

Honor the Earth
Native-led support for grassroots efforts; forging change in Native American country. honorearth.org

Human Development Reports
New York, NY; Research, analysis, advocacy for new ideas and policies to promote opportunities advancing human development. hdr.undp.org

Human Rights Watch
New York, NY; Brings attention to human rights conditions around the world and works for change in policy and practice. hrw.org

Humanitad
Ubud, Bali, Indonesia; Promotes interfaith and intercultural fellowship through global arts, culture and education. humanitad.org

Hunger Project
New York, NY; Supports grassroots, women-centered strategies for sustainable, self-reliant solutions to hunger around the world. thp.org

Immigration Policy Center
Washington, DC; Accurate information about the effects of immigration on the US economy and society. immigrationpolicy.org

Indigenous Environmental Network
Bemidji, MN; Indigenous Peoples of America working from a grassroots respect of tradition and natural laws to protect lands from contamination and exploitation. ienearth.org

Indigenous Women's Network
Austin, TX; Supports indigenous women around the world and the next generation to work proactively for sustainable communities. indigenouswomen.org

Innovations in Civic Participation
Washington, DC; Promotes sustainable development and social change through youth service and learning around the world. icicp.org

International Labour Organization; Indigenous and Tribal Peoples
Geneva, Switzerland; Advocates for social justice and human and labor rights for Indigenous and Tribal Peoples around the world. ilo.org/indigenous

Environmental Service Learning Initiative (eslisf.org)

League of United Latin American Council
Washington, DC; Serving Hispanics with a full range of community-based programs; operating 700 councils across the US. lulac.org

Move to Amend
US nationwide; Committed to social and economic justice, ending corporate rule, and building a vibrant democracy. movetoamend.org

National Conference for Community and Justice
Chapters around the US; Advocates respect for all races and religions through conflict resolution and education. nccj.org

National Council of La Raza
Largest Latino civil rights and advocacy in the US; research, policy analysis and support for the Latino perspective. nclr.org

National Network for Immigrant and Refugee Rights
Oakland, CA; Informs, trains and acts in support of equal rights issues for immigrants and refugees. nnirr.org

Natural Resource Conservation Service
Washington, DC; Aids private landowners in their stewardship of healthy ecosystems. nrcs.usda.gov

Navdanya International
New Delhi, India; Supports small sustainable farming with conservation, biodiversity and the conservation of crops and plants near extinction. navdanya.org

ONE
Worldwide; Policy and trade reform campaigns against disease and poverty worldwide, especially Africa. one.org

Oxfam International
Oxford, UK; Global organization working to end poverty through assistance, advocacy and policy research. oxfam.org

Pew Hispanic Center
Pew Global Attitudes Project; research, news, and trends about Hispanics and Latinos. hispanicpewresearch.org

Population Connection
Washington, DC; U.S. grassroots organization that educates and advocates for sustainable population planning and policy. populationconnection.org

Population Institute
Washington, DC; Seeks to reduce excessive population growth through universal access to family planning, education, and services. populationinstitute.com

Population Reference Bureau
Washington, DC; Empowering information to advance the wellbeing of current and future generations. prb.org

REACH Center
Seattle, WA; Education and training about and for multiculturalism and diversity. reachctr.org

Rockwood Leadership Program
Oakland, CA; Promotes social change through training progressive nonprofits. rockwoodfund.org

Roots of Peace
San Rafael, CA; Removing land mines in exchange for vines. rootsofpeace.org

Ruckus Society
Oakland, CA; Nonviolent training for social and environmental justice organizations. ruckus.org

Seventh Generation Fund for Indian Development
Upholding the sovereignty of Native peoples, with grassroots support. 7genfund.org

Southern Poverty Law Center
Montgomery, AL; Civil rights/anti-hate group law center, promotes tolerance. splcenter.org

Survival International
UK; Working for tribal peoples' rights through education, advocacy and campaigns. survival-international.org

Terra Madre
Bra (CN) Italy; International projects supporting sustainable local food production, through Slow Food communities network. terramadre.info

The Pachamama Alliance
San Francisco, CA; Empowering indigenous rainforest people, promoting new global vision. pachamama.org

United for a Fair Economy
Boston, MA; Popular economics education bridging the equity divide; great library, training tools. faireconomy.org

United Nations Association (USA)
New York, NY; UNA-USA encourages responsible US leadership in the United Nations. unausa.org

United Nations Development Programme
New York, NY; Helps developing countries build sustainable communities, equal rights. undp.org

United Students Against Sweatshops
Washington, DC; Students organizing for sweatshop-free labor conditions and workers' rights. usas.org

Universal Living Wage
Austin, TX; Campaign to make the Federal Minimum Wage correlated to affordable housing, food and clothing. universallivingwage.org

US Labor Education in the Americas Project
Supports economic justice and rights for workers in Latin America. usleap.org

Via Campesina
Honduras, Zimbabwe; Coordinates small- and middle-scale producers, agricultural workers, rural women, and indigenous communities. viacampesina.org

White Earth Land Recovery Project
Native Harvest catalog, alternative energy projects, preserving cultural heritage. nativeharvest.com

Witness for Peace
Washington, DC; Campaigns for social justice and sustainable economies in Latin America and the Caribbean. witnessforpeace.org

Women for Women International
Washington, DC; Addresses the unique needs of women in conflict and post-conflict environments. womenforwomen.org

Women's Learning Partnership
Bethesda, MD; Empowering women to be leaders in transforming their families and communities, particularly in Muslim-majority societies. learningpartnership.org

World Vision International
Christian group that provides emergency aid, long-term sustainability development resources, and campaigns for policy reform. wvi.org

ECONOMICS

Accion
Boston, MA; Poverty alleviation through micro loans, training, financial services. accion.org

Ad Busters
Global network of culture jammers and creatives working to change the way meaning is produced in our society. adbusters.org

American Booksellers Association
Tarrytown, NY; Promotion and advocacy for independent booksellers nationwide. bookweb.org

American Independent Business Alliance (AMIBA)
Bozeman, MT; Promotes community-based independent businesses. amiba.net

American Pictures
Denmark; A multi-media presentation on racism, oppression, and the underclass. american-pictures.com

Association for the Taxation of Financial Transactions for the Aid of Citizens
Worldwide; Advocating the Tobin tax, and democratic control of global financial markets/ institutions. attac.org

B Corp
Wayne, PA; Campaigning for businesses to become socially responsible through a ratings system, laws and collaboration. bcorporation.net

Bank Information Center
Washington, DC; Advocates social and economic justice through world financial institutions to support ecological sustainability in developing countries. bicusa.org

Bay Localize
Oakland, CA; Catalyzing regional self-reliant economy for SF Bay Area, providing localize toolkits for communities. baylocalize.org

Better Business Bureau
Arlington, VA; Sets standards and supports best business practices for a trustworthy marketplace. bbb.org

Better World Club
Portland, OR; USA's only environmentally friendly auto club. betterworldclub.com

Biomimicry 3.8
Missoula, MT; Promotes sustainable, energy efficient, high performance design strategy for products and processes. biomimicryguild.com

Business Alliance for Local Living Economies
Oakland, CA; Conferences, training, and tools for localization of wealth. bealocalist.org

Business for Innovative Climate and Energy Policy
Boston, MA; Works with businesses and policy makers to support legislation that counters the risks of climate change. ceres.org/bicep

Calvert Foundation
Bethesda, MD; Affordable capital for needy individuals, communities and nonprofits worldwide. calvertfoundation.org

Canadian Co-operative Association
Canada; Represents coop and credit union members in Canada and 40 countries. coopscanada.coop

Center for the Advancement of the Steady State Economy
Arlington, VA; Promotes the steady state economy as a desirable alternative to economic growth. steadystate.org

Center for American Progress
Washington, DC; Develops progressive ideas and actions, and challenges conservative misinformation. americanprogress.org

Center for Community Self-Help
Durham, NC; Credit union; helps borrowers nationwide, particularly the vulnerable population. self-help.org

Center for Partership Studies
Pacific Grove, CA; Promotes realization of human capacities for consciousness, caring, and creativity. partnershipway.org

Center for Popular Economics
Amherst, MA; Organization of economists offering resources to help build a sustainable and just economy. populareconomics.org

Center for State Innovation
Madison, WI; Helping states implement innovative, progressive policies. stateinnovation.org

Center for Sustainable Economy
Santa Fe, NM; Develops and supports political and economic campaigns promoting a sustainable society. sustainable-economy.org

Center of Concern
Catholic research, educational org promoting just, sustainable international finance, trade systems. coc.org

CERES
Boston, MA; Investors, environmental, public interest groups working with corporations to address climate change. ceres.org

ChevronToxico
San Francisco, CA; Amazon Watch campaign to pressure Chevron/Texico to clean up hazardous oil waste dumped in Ecuador. chevrontoxico.com

Citizens Trade Campaign
Washington, DC; National coalition opposing NAFTA; promoting just trade legislation in US. citizenstrade.org

Common Frontiers
Ontario, Canada; Examining the effects of economic integration in the Americas. commonfrontiers.ca

Community Environmental Legal Defense Fund
Mercersburg, PA; Legal services for building sustainable communities and ending corporate governance. celdf.org

Community Investment Network
Durham, NC; Finances, summits, training for community empowerment. thecommunityinvestment.org

CooperationWorks!
Hanover, MN; Resources to help communities develop cooperative economies. cooperationworks.coop

Cooperative Development Services
Madison, WI; Organizational and development guidance to start and expand co-ops and producer-owned businesses. cdsus.coop

Corporate Responsibility Coalition
London; UK; Alliance promoting respect for rights of workers, local communities and the environment. corporate-responsibility.org

CorpWatch
San Francisco, CA; Investigative journalism to support campaigns against corporate abuses in communities worldwide. corpwatch.org

Cost of War
Northampton, MA; Running calculator of the costs of war. costofwar.com

Council of Canadians
W. Ottawa, Canada; Citizens' watchdog org promoting economic justice, environmental preservation. canadians.org

Economic Policy Institute
Washington, DC; Broadening public debate strategies to achieve prosperous and fair economy. epi.org

Ethical Markets
St. Augustine, FL; Provides analysis on green markets and innovations. ethicalmarkets.com

Fair Trade Federation
Wilmington, DE; Global movement in support of local businesses and trade. fairtradefederation.org

Fair Trade International
Bonn, Germany; Fair trade strategies to support sustainable livelihoods for disadvantaged producers. fairtrade.net

Fair World Project
Portland, OR; Expands markets for authentic fair trade, educates consumers about issues in trade and agriculture. fairworldproject.org

FINCA
Washington, DC; Financial loans to low income entrepreneurs in the world, particularly women, with village support systems. finca.org

Forest Stewardship Council
Minneapolis, MN; Supports sustainable forest management practices and promotes certified forest products. fsc.org

Global Environmental Mgmt. Initiative
Washington, DC; Guidebook on forging corporate-nonprofit partnerships. gemi.org

Global Giving
Washington, DC; Research about donations to prescreened worldwide grassroots organizations. globalgiving.com

Global Greengrants Fund
Boulder, CO; Connects donors with grants for grassroots efforts of sustainability around the world. greengrants.org

Global Policy Forum
New York, NY & Bonn, Germany; Monitors UN policy, informs citizen action for peace and justice. globalpolicy.org

Grameen Bank
Bangladesh; Providing micro credit to the poor in rural Bangladesh. grameen-info.org

Grassroots.org
W. Jordan, UT; Serving nonprofits with free information, internet services, legal consulting. grassroots.org

Grassroots Leadership
Charlotte, NC; Training and resource center to end social and economic oppression. grassrootsleadership.org

Green America
Washington, DC; Providing economic solutions via National Green Pages, resources, Green Festivals and more. greenamerica.org

Green Biz
Oakland, CA; Provides resources to businesses to enable environmental responsibility along with profitability. greenbiz.com

Greenpeace Greenwashing
Washington, DC; Corporate watchdog, accountability advocate, exposing deceptive environmental and social claims and ads. stopgreenwash.org

Hemispheric Social Alliance
Bogota, Colombia; Organizations from the Americas coordinated to promote economic and social justice. asc-hsa.org

Interfaith Center on Corporate Responsibility
New York, NY; Religious and social movements working collaboratively for social, economic and environmental sustainability. iccr.org

International Co-operative Alliance
Geneva, Switzerland; Uniting, representing and serving co-operatives worldwide. ica.coop

International Forum on Globalization
San Francisco, CA; Stimulating new thinking and responses to economic globalization. ifg.org

International Society for Ecology and Culture
UK, US, Germany; Promotes locally-based alternatives to global consumerism. localfutures.org

Investors' Circle Foundation
San Francisco, CA & Durham, NC; Connecting investors with startup businesses committed to confronting social and environmental challenges. investorscircle.net

Jobs with Justice
Washington, DC; Organizes and supports community campaigns for workplace equitability and connection with the local community. jwj.org

Jubilee USA Network
Washington, DC; Campaigning to cancel debts of poorer nations and promote responsible and just financing by the US. jubileeusa.org

Kiva
San Francisco, CA; Web-based microlending to alleviate poverty around the world. kiva.org

Living Economies Forum
Bainbridge Island, WA; Strategies and mindsets that support biodiverse, self-sustainable, bottom-up economies. livingeconomiesforum.org

Natural Capitalism Solutions
Longmont, CO; Guides government agencies and corporations toward equitable, sustainable and efficient practices. natcapsolutions.org

Natural Logic
Berkeley, CA; Advises businesses on profitability along with sustainability and social responsibility. natlogic.com

Natural Step
Sweden and Portland, OR; Resources and guidance for organizations and businesses to adapt principles of world sustainability. naturalstep.org

New Economics Foundation
London, UK; Challenging mainstream thinking on economic, environmental, social issues. neweconomics.org

New Economy Coalition
Boston, MA; Works to build a New Economy that prioritizes the wellbeing of people and the planet. neweconomy.net

New Economy Working Group
Washington, DC; Virtual think tank reframing the economic policy debate. neweconomyworkinggroup.org

Northwest Earth Institute
Portland, OR; Courses and materials for initiating community discussion about sustainability. nwei.org

Oakland Institute
Oakland, CA; Influential international think tank regarding social, economic and environmental issues. oaklandinstitute.org

Right to the City
Organizes community campaigns for social and economic justice. righttothecity.org

Slow Money
Boulder, CO; Promotes economic ventures by connecting investors with local, organic food enterprises. slowmoney.org

Soul of Money Institute
San Francisco, CA; Workshops and publications to help individuals and organizations effectively use their finances. soulofmoney.org

SourceWatch
Madison, WI; Nonprofit investigative reporting group, exposing people and groups shaping public policy. sourcewatch.org

Teach a Man to Fish
London, UK; Supports schools, education and vocational training to achieve self-sufficiency. teachamantofish.org.uk

Third World Network
Penang, Malaysia; Provides info to educate and advocate for sustainable and just development in developing countries. twnside.org.sg

Transnational Institute
Netherlands; Network of activist-scholars providing critical analyses of global problems. tni.org

UN-HABITAT
Nairobi, Kenya; United Nations program of research, financing, advocacy for sustainable and fair urban development. unhabitat.org

UN Global Compact
New York, NY; Resources for businesses in sup-port of sustainable society. unglobalcompact.org

World Council of Credit Unions
Advocates for credit unions; information and serivces about finance. woccu.org

World Fair Trade Organization
The Netherlands; Support and advocacy for the small producers worldwide. wfto.com

WorldChanging
Seattle, WA; Open source network of many solutions and innovations. worldchanging.org

World Resources Institute
Washington, DC; Provides research and strategizing for natural resources throughout the world. wri.org

Art Explosion

COMMUNITIES

Action Without Borders
Portland, OR & New York, NY; Connecting people w/resources to help build a free and dignified world. idealist.org

American Community Gardening Association
US, Canada; Resources, support and networking for community gardening. communitygarden.org

American Jewish World Service
New York, NY; Supporting communicating-based development to alleviate poverty, hunger, disease throughout the world. ajws.org

Business Alliance for Local Living Economies
Oakland, CA; Educates, connects leaders, spreads solutions, and drives investment toward local economies. bealocalist.org

Bay Area Green Tours
Berkeley, CA; Provides educational tours and events that show the sustainable economy in action and inspires support of local green businesses. bayareagreentours.org

Be the Change Earth Alliance
Vancouver, BC, Canada; Support for collective and individual work towards a healthy, sustainable world. bethechangeearthalliance.org

Berkeley Community Gardening Collaborative
Berkeley, CA; Programs supporting community sustainability and a healthy planet. ecologycenter.org/bcgc

Brower Youth Awards
Berkeley, CA; Promoting youth environmental leadership. broweryouthawards.org

Building Green
Brattleboro, VT; Provides resources and services for environmentally-conscious projects. buildinggreen.com

Care Action Network
Atlanta, GA; Humanitarians fighting global poverty, putting women at center of community efforts. care.org

Center for Neighborhood Technology
Chicago, IL; Think tank of research and strategies to promote urban sustainability. cnt.org

Co-Intelligence Institute
Eugene, OR; Promoting community dialogue on how to tap wisdom together. co-intelligence.org

CommonDreams.org
Portland, ME; Nonprofit, progressive and independent news center. commondreams.org

Community Solutions
Yellow Springs, OH; Survival strategies re: peak oil and climate change. communitysolution.org

Creative Education Foundation
Scituate, MA; Programs for leaders of innovation aiding improvement in societal function. creativeeducationfoundation.org

David Suzuki Foundation
Vancouver, Canada; Conserving environment through science-based education, advocacy, policy work. davidsuzuki.org

Dictionary of Sustainable Management
San Francisco, CA; Open dictionary of sustainability terms. sustainabilitydictionary.com

Donella Meadows Institute
Norwich, VT; Resources for a sustainable world, and to develop northern New England as a model region. donellameadows.org

Dynamic Facilitation
Port Townsend, WA; For effective meetings, resolving conflicts, co-creating solutions and fostering true democracy. tobe.net

Earth Charter Initiative
San José, Costa Rica; Promotes ethical principles for building a just, sustainable, peaceful global society. earthcharterinaction.org

Earthaven
Black Mountain, NC; Intentional community offering short term experiential sustainable living training. earthaven.org

Eco-Logica
Lancaster, UK; Publications, resources and commissions for sustainability through accessibility and transportation planning. eco-logica.co.uk

Big Stock Photo

Ecology Center
Berkeley, CA; Sustainable living models, tools, training, referrals, strategies, infrastructure. ecologycenter.org

Enclude
Washington, DC; Connecting entrepreneurs with capital and advisory resources. encludesolutions.org

Energy Star
Washington, DC; US Environmental Protection Agency and Dept. of Energy promote energy efficiency. energystar.gov

Equality Now
New York, NY; Working to end violence, discrimination against women, girls, globally. equalitynow.org

Fair Vote
Takoma Park, MD; Transforming elections for universal access to participation, and a full spectrum of ballot choices. fairvote.org

Farm Animal Rights Movement
Bethesda, MD; Advocating plant-based diet, humane treatment of farm animals. farmusa.org

Food Not Bombs
Arroyo Seco, NM; Working to end hunger and stop economic and environmental exploitation. foodnotbombs.net

Girls Learn International
Arlington, VA; Pairs US classrooms with spartan classrooms in Africa, Asia, Latin America. girlslearn.org

Global Ecovillage Network
Worldwide; Promoting and fostering international sharing of ideas for sustainable communities. gen.ecovillage.org

Global Fund for Women
San Francisco, CA; Fosters empowerment and health for women worldwide with funding, education and mobilized resources. globalfundforwomen.org

Global Green USA
Santa Monica, CA; Advocates for solutions to global warming through education, policies, finances and support of low-income communities. globalgreen.org

Global Giving
Washington, DC; Fundraising site for social entrepreneurs and nonprofits worldwide who are improving their communities. globalgiving.org

Global Oneness Project
Inverness, CA; Films, articles, photography used to educate for environmental and multicultural issues. globalonenessproject.org

Good Works
Washington, DC; National directory of over 1000 social change corporations. goodworksfirst.org

Green America
Washington, DC; Promotes social responsibility through many programs that educate individuals and groups. greenamerica.org

Green Cross International
Geneva, Switzerland; Cooperative intl. projects to seek and provide paths towards sustainable, peaceful and healthy communities. gci.ch

Green Festivals
San Francisco, CA; Green products and services annually showcased in several major U.S. cities. greenfestivals.com

GreenerCars
Washington, DC; Rating the environmental friendliness of every vehicle on the market. greenercars.com

Grist
Seattle, WA; Environmental news written to promote discussion and action. grist.org

Habitat for Humanity
International; Builds and repairs houses around the world with volunteers, donations and affordable loans. habitat.org

Haute Couleur
Forum for artists to display their art and support the global community with profit donations. hautecouleur.org

Healthy Built Homes
NC; Resources for eco-friendly home materials and contractors. healthybuilthomes.org

The Heritage Institute
Freeland, WA; On-line, on-site continuing sustainability education for teachers. hol.edu

Housing and Urban Development
Washington, DC; Federal support for affordable housing. hud.gov

Institute for Sustainable Communities
Montpelier, VT; Trains and mentors communities in sustainability. iscvt.org

Institute for Transportation and Development Policy
New York, NY; Promoting environmentally sustainable and socially equitable transportation worldwide. itdp.org

Institute of Noetic Sciences
Petaluma, CA; Research on the potential and power of consciousness. noetic.org

Land Trust Alliance
Washington, DC; Works for conservation of national land through policy. landtrustalliance.org

NextGEN
Youth working towards sustainability and stronger communities. nextgen.ecovillage.org

Occidental Arts and Ecology Center
Occidental, CA; Programs on permaculture, hydrology and bioremediation. oaec.org

Omega Institute for Holistic Studies
Rhinebeck, NY; Offers educational workshops and retreats on living holistically. eomega.org

One World
London, UK; Supports development of educational Internet and mobile phone apps for the world's poorest people. oneworld.org

Open Space Technology
Resources for effective collaborative conferencing and problem solving. openspaceworld.org

Partners for Livable Communities
Washington, DC; Leadership organization to assist community development. livable.com

Pathfinder International
Watertown, MA; Working in Africa, Asia and South America to provide quality sexual health care. pathfind.org

Peace Action
Silver Spring, MD; Grassroots national affiliates mobilized for progressive political strategies against war and violence. peace-action.org

Peace Alliance
Washington, DC; Education and activism to make peace central to society. peacealliance.org

Plan USA
Washington, DC; Grassroots programs in health, education, water, sanitation, income-generation. planusa.org

Planetwork
San Francisco, CA; Digital technology for democratic, socially just, ecologically sane future. planetwork.net

Project for Public Spaces
New York, NY; Planning, designing and educating for sustainable cities worldwide. pps.org

Reconnecting America
San Francisco, CA; Symposiums and online learning and collaboration for action empowerment to improve the world. uptous.org

Public Radio Capital
Boulder, CO; Strategizes to promote accessibility and capital for public radio stations. publicradiocapital.org

Right Livelihood Award
Stockholm, Sweden; "Alternative Nobel prize" given annually to pioneers in creating a just and sustainable future. rightlivelihood.org

Save the Children
Westport, CT; Creating lasting, positive change for children in need. savethechildren.org

Silicon Valley Toxics Coalition
San Francisco, CA; Coalition concerned with health problems caused by high-tech electronics industry. svtc.org

Small Planet Institute
Cambridge, MA; Investigating root causes of global issues and suggestions for how individuals can help. smallplanetinstitute.org

Sustainable Communities Online
Washington, DC; Resources for building integrative and sustainable communities. sustainable.org

Tapestry Institute
Longmont, CO; Research, scholarship and education emphasizing kinship with the Earth. tapestryinstitute.org

Transition Network
Totnes, UK; Provides resources to support communities worldwide becoming sustainable and resilient. transitionnetwork.org

Transition US
Sebastopol, CA; Provides resources to support communities worldwide becoming sustainable and resilient. transitionus.org

Tree Hugger
Atlanta, GA; Media outlet; clearing house for green news, solutions, products. treehugger.com

US Green Building Council
Washington, DC; Advocates for green building practices through education, conferencing, mobilization and policies. usgbc.org

Union of Concerned Scientists
Cambridge, MA; Works towards securing changes in policy, corporate practices, consumer choices. ucsusa.org

Urban Land Institute
Washington, DC; Research, education and advocacy for sustainable real estate development and land use policies. uli.org

Urban Permaculture Guild
Oakland, CA; Inspiring communities to creatively transform how and where they live. urbanpermacultureguild.org

Venus Project
Venus, FL; Research and ideas for futuristic sustainable world models. thevenusproject.com

Weatherization Assistance Program
Federal program assisting low-income families improve energy efficiency of homes. waptac.org

What's Your Tree?
Creating an intl. network of small groups that work to improve the world. whatsyourtree.org

WiserEarth
San Francisco, CA; Connecting people, nonprofits, businesses for a just and sustainable world. naturalcapital.org/wiser.htm

Women for Women
Washington, DC; Helping women survivors of war gain stability and self-sufficiency. womenforwomen.org

Women's Earth Alliance
Berkeley, CA; Educates and supports women in communities worldwide to help find solutions for water, food, land and climate change issues. womensearthalliance.org

World Green Building Council
Transforming building practices for thriving environment, economy and society. worldgbc.org

Yes! Magazine
Bainbridge Island, WA; Magazine with emphasis on empowering people to work for a healthy planet. yesmagazine.org

YES! Youth for Environmental Sanity
Soquel, CA; Mobilizing young adults around the world into collaborative efforts for community building and healing. yesworld.org

GETTING PERSONAL

Action for Solidarity, Equality, Environment, and Development (A SEED)
The Netherlands; Mobilizes youth for campaigns against environmental destruction and social injustice. aseed.net

Agape International Spiritual Center
Culver City, CA; Active teaching and practice of the New Thought-Ancient Wisdom tradition of spirituality. agapelive.com

Alliance for a New Humanity
San Juan, Puerto Rico; Personal and social transformation to build a just, peaceful, and sustainable world. anhglobal.org

Awakening the Dreamer, Changing the Dream Symposium
San Francisco, CA; Inspiring an environmentally sustainable, spiritually fulfilling, socially just human presence. uptous.org

Better World Handbook
Davis, CA; Tips for green living; excellent sustainability overview. betterworldhandbook.com

Better World Shopping Guide
Davis, CA; 1000+ company database for eco-conscious consumers. betterworldshopper.org

Beyond Pesticides
Washington, DC; Provides information promoting protection of public health and the environment. beyondpesticides.org

Big Picture Small World
Media, PA; Education to empower student leaders. bigpicturesmallworld.com

Biodegradable Products Institute
New York, NY; Professional association that educates and sets standards for the use of recyclable materials. bpiworld.org

Bioneers
Santa Fe, NM; Creates conferences, publications, and broadcasts of innovators sharing solutions to global problems. bioneers.org

CarSharing Association
Illinois; Promotes and educates regarding the car sharing industry and has member organizations around the world. carsharing.org

Care2
Redwood City, CA; Social action network providing information and opportunities for action. care2.com

Center for a New American Dream
Charlottesville, VA; Raises awareness of connection between consumerism and effects on the environment. newdream.org

Center for Food Safety
Washington, DC; Promotes organic and sustainable agriculture. centerforfoodsafety.org

Center for Teen Empowerment
Roxbury, MA; Inspires young people to handle difficult social problems, create positive change. teenempowerment.org

Center for Visionary Leadership
San Rafael, CA; Training to promote visionary leadership for social change through business and politics. visionarylead.org

Challenge Day - Youth
Concord, CA; Workshops on school campuses and training for adult leaders to promote teen empowerment. challengeday.org

Children of the Earth
South Burlington, Vermont; Programs for global youth to foster leadership skills and cooperation. coeworld.org

Circle of Life
Oakland, CA; Provides education and inspiration to live in a way that honors diversity and interdependence of all life. circleoflife.org

Common Cause
Washington, DC; Lobbying, education, grassroots efforts and press outreach to make politicians accountable and transparent. commoncause.org

Daily Acts
Petaluma, CA; Inspiring choices that matter; sustainability education. dailyacts.org

Daily Green
Port Ewen, NY; Online news, information and blogging about green living. thedailygreen.com

Dream Change
Bainbridge Island, WA; Webinars, workshops, and conferences that educate to live in a way that ensures a healthy world. dreamchange.org

Ecological Footprint
Ecological Footprint Quiz: estimate your daily use of earth resources. myfootprint.org

Empowerment Institute
West Hurley, NY; Research, training and publications for empowering organizations and individuals to transform themselves and their communities. empowermentinstitute.net

Roberta Vogel

Evangelical Environmental Network
New Freedom, PA; Christian resources for improving the environment around the world. creationcare.org

Focus the Nation
Portland, OR; Training students to work for clean energy in their communities and on their college campuses. focusthenation.org

Food First
Oakland, CA; Provides research, education and support for communities and projects of equitable and healthy food systems around the world. foodfirst.org

Forum for Food Sovereignty
A global network of NGOs/CSOs concerned with food sovereignty issues. foodsovereignty.org

Foundation for Conscious Evolution
Santa Barbara, CA; Supporting the awakening of the spiritual, social, and scientific potential of humanity. barbaramarxhubbard.com

Funders' Collaborative on Youth Organizing
Brooklyn, NY; Provides resources supporting youth-oriented efforts to build healthy and equitable communities. fcyo.org

Generation Waking Up
Oakland, CA; Training for youth to become active in building a sustainable world. generationwakingup.org

Global Youth Action Network
Accra, Ghana; A collaboration of youth problem-solving organizations worldwide. gyan.tigweb.org

Global Youth Connect
Empowering youth to advance human rights and create a more just world. globalyouthconnect.org

Gratefulness.org
Ithaca, NY; Promotes gratefulness as a common language for all people and a path to living peacefully and sustainably. gratefulness.org

Green Guide for Everyday Living
New York, NY; Resources, articles and tips for green living. thegreenguide.com

Initiative on Children's Environmental Health
Eliminating environmental exposures that undermine health. iceh.org

Imago Relationships
Lexington, KY; Transforms relationships and the world through empathic dialogue. gettingtheloveyouwant.com

Institute of HeartMath
Boulder Creek, CA; Research, education, training and resources for improving ability to lead a healthy life. heartmath.org

Institute for Responsible Technology
Fairfield, IA; Features the Non-GMO Shopping Guide. responsibletechnology.org

IPAS
Chapel Hill, NC; Mobilizes training and education for safe abortions in the poorest countries. notyetrain.org

Landmark Education
San Francisco, CA; A powerful, accelerated learning experience affecting quality of life. landmarkeducation.com

Lifestyles of Health and Sustainability
Louisville, CA; Green living tips for businesses and consumers. lohas.com

Matrona
Asheville, NC; Promoting midwifery and undisturbed birth. thematrona.com

Move On
Berkeley, CA; Advocates grassroots campaigns for social change; huge membership. moveon.org

National Sustainable Agriculture Coalition
Washington, DC; Coalition of farmers, environmentalists, and consumer advocates. sustainableagriculture.net

Next Generation
San Anselmo, CA; Promotes social empowerment in youth through educational programs about sustainability, social justice and peace. gonextgeneration.org

One World Now - Youth
Seattle, WA; Providing global leadership opportunitites for youth. oneworldnow.org

Organic Consumers Association
Finland, MN; Mobilization of grassroots actions to support fair trade, local agriculture, and sustainable practices. organicconsumers.org

Public Citizen
Washington DC & Austin, TX; Founded by Ralph Nader; consumer protection and government accountability. citizen.org

Regenerative Design Institute
Bolinas, CA; Teaching dynamic permaculture skills necessary for sustainable living. regenerativedesign.org

Resources for Spirituality Journeys
Spotlights people of different religious, spiritual traditions. spiritualityandpractice.com

Roots and Shoots
Vienna, VA; Provides young people with the knowledge, tools and inspiration to improve the environment and quality of life for people and animals. rootsandshoots.org

Square Foot Gardening
Columbia, SC; Natural, high-yield garden method using less space, and water than conventional methods. squarefootgardening.com

Story of Stuff
Berkeley, CA; Grassroots efforts to promote sustainability and a just economy with solutions to unhealthy consumerism. storyofstuff.com

Up to Us
San Francisco, CA; Symposiums and education to mobilize action and consciousness for a sustainable and just world. uptous.org

What Kids Can Do
Providence, RI; Promotes the power of global youth through projects, organizations and collaboration. whatkidscando.org

Wisdom of the World
Greenbrae, CA; Produces musical solutions that support people to face their life transitions in an emotionally sustainable way. wisdomoftheworld.com

Big Stock Photo

ENDNOTES

Part One: SETTING THE CONTEXT

Quotes and figures sourced from The Pachamama Alliance website, www.pachamama.org.

1. Symposium website: http://www.awakeningthedreamer.org

2. http://www.joannamacy.net/theworkthatreconnects/goals.html

3. Earth under severe stress; stats from sites listed below: ozone: http://www.livescience.com/environment/080612-ozone-warming.html
   forest: http://www.un.org/earthwatch/forests/forestloss.html
   land: http://www.worldometers.info/ and http://www.fao.org/docrep/t0667e/t0667e04.htm
   fish: http://overfishing.org

4. UN, Section 3, #27 of Our Common Future, 1987: http://www.un-documents.net/wced-ocf.htm

5. Charts and stats available on http://www.footprintnetwork.org

6. The World Wildlife Fund's Living Planet Index: http://assets.wwf.org.uk/downloads/lpr_2012_rio_summary_booklet_final_9may2012.pdf

7. If the Earth had 100 people, ratio description: http://miniature-earth.com

8. The World Bank: http://web.worldbank.org/WBSITE/EXTERNAL/EXTABOUTUS/0,,contentMDK:23272497~pagePK:51123644~piPK:329829~theSitePK:29708,00.html?argument=value

9. Stats relayed in larger context at http://www.globalissues.org/TradeRelated/Facts.asp

10. http://www.youtube.com/watch?v=9GorqroigqM

11. http://www.youtube.com/watch?v=7DLokOQZlag

Part Two: ENVIRONMENT

1. Center for Climate and Energy Solutions: http://www.c2es.org/docUploads/climate101-overview.pdf

2. The Pew Center, "Mythbusters" pdf: www.pewclimate.org

3. http://www.epa.gov/rlep/faq.html and http://www.epa.gov/climatechange/ghgemissions/sources/agriculture.html

4. http://www.humanesociety.org/assets/pdfs/farm/hsus-fact-sheet-greenhouse-gas-emissions-from-animal-agriculture.pdf

5. http://www.ucsusa.org/global_warming/science_and_impacts/science/findings-of-the-ipcc-fourth-2.html

6. http://www.ncdc.noaa.gov/sotc/national/2013/11/supplemental/page-4/ about "unusually wet conditions in the Midwestern and eastern US"

Sidebar

(a) (Terminology for the 21st Century): The Pew Center for Global Climate Change, www.pewclimate.org; The Intergovernmental Panel on Climate Change, www.ipcc.ch; International Rivers, www.internationalrivers.org

7. http://citizensclimatelobby.org/video-about-citizens-climate-lobby/

8. Science Daily (Aug 13, 08): www.sciencedaily.com/releases/2008/08/080813144405.htm

9. Worldwatch Institute: State of the World 2009

10. http://frontierscientists.com/2012/10/ocean-acidification/

11. http://www.ocean-acidification.net/FAQeco.html

12. Worldwatch Institute: State of the World 2008

13. http://www.fao.org/docrep/016/i2727e/i2727e.pdf

14. http://www.epa.gov/owow/nps/qa.html

15. The Center for Biological Diversity: www.biologicaldiversity.org

16. http://www.fao.org/docrep/016/i2727e/i2727e.pdf

17. http://www.wri.org/press/2011/02/press-release-75-worlds-coral-reefs-currently-under-threat-new-analysis-finds

18. http://www.gcrmn.org/pdf/P1%20Status%202009%20Brochure%207th%20draft.pdf

19. http://www.icriforum.org/about-icri and http://www.conservation.org/learn/biodiversity/species/profiles/corals/Pages/corals.aspx

20. http://kurrawa.gbrmpa.gov.au/corp_site/info_services/publications/sotr/ and https://www.getup.org.au/campaigns/coal-seam-gas/great-barrier-reef/save-the-reef

21. The Species Alliance: http://speciesalliance.org

22. International Rivers: http://www.internationalrivers.org

23. http://www.un.org/millenniumgoals/pdf/MDG%20Report%202012.pdf

24. Public Citizen, http://www.citizen.org and www.wateractivist.org; Food & Water Watch, http://www.foodandwaterwatch.org; Occidental Arts and Ecology Center, http://www.oaec.org

Sidebar

(b) (True Thirst): Statistics from Jean-Michel Cousteau, "Seas the Day: Water is World's Key Security Issue," Ocean Futures Society, June 2004 feature story, http://www.oceanfutures.org

25, 26. Food & Water Watch: http://www.foodandwaterwatch.org

Sidebar

(c) (Bottle Water Consumption): http://www.washingtonpost.com/wp-dyn/content/article/2005/09/19/AR2005091901295.html

27. USDA-FS 2000, accessed Sept 09 on www.ran.org, site of the Rainforest Action Network

28. http://www.ran.org

29. Elisabeth Rosenthal, "In Brazil, Paying Farmers to Let the Trees Stand," The New York Times (Aug 22, 09), front page

30. UN Global Forest Resource Assessment 2005: http://www.ran.org

31. Marc Gunther, "Eco-police find new target: Oreos," Fortune magazine (Aug 13, 08)

32. http://wwf.panda.org/what_we_do/footprint/agriculture/soy/soyreport/

33. Center for Biological Diversity: http://www.biologicaldiversity.org

34. E.O. Wilson, The Creation: An Appeal to Save Life on Earth (2006)

35. Extinction research by professor David Ulansey: http://www.well.com/~davidu/extinction.html

36. Trish Andryszewski, Mass Extinction (2008)

37. International Union for the Conservation of Nature (IUCN), 2009, Red List, sourced on http://www.biologicaldiversity.org

38. Trish Andryszewski, Mass Extinction, see note 38

39. http://www.iss.it/binary/publ/cont/ANN_08_04%20Binetti.1209032191.pdf

40. Trish Andryszewski, Mass Extinction, see note 38

41. Algalita Marine Research Foundation: http://www.algalita.org/pelagic_plastic.html

42, 43. Brian Halwell, "Ocean Pollution Worsens and Spreads," see note 33

Part Three: ENERGY

Sidebar

(a) (Energy Opportunities to Reduce Global Warming): NRDC, Fig 1 in http://docs.nrdc.org/nuclear/nuc_08042301A.pdf

1. Christopher Flavin, Worldwatch Report: "Low-Carbon Energy—A Roadmap," Renewable Energy

2. U.S. Energy Information Administration: http://www.eia.gov/tools/faqs/faq.cfm?id=527&t=1

3. http://www.instituteforenergyresearch.org/energy-overview/solar/

4. NRDC: http://www.nrdc.org/air/energy/renewables/solar.asp

5. http://www.nrdc.org/energy/renewables/wind.asp#footnote2

6. http://www.hcn.org/issues/45.22/the-latest-first-federal-prosecution-of-wind-farm-bird-deaths and http://www.motherjones.com/environment/2014/01/birds-bats-wind-turbines-deadly-collisions

7. http://www.worldwatch.org/node/5748

8. http://energy.gov/oe/technology-development/smart-grid

9. US Geological Survey (USGS), "Geothermal Energy: Clean Power from the Earth's Heat," accessed Aug 08 at http://www.nonukes.org/library/usgs_2003_report_on_geothermal_energy.pdf

10. http://www1.eere.energy.gov/geothermal/faqs.html

11. http://environment.nationalgeographic.com/environment/global-warming/geothermal-profile/

12. USGS, see note 7

Sidebar

(b) (Renewable Energy DIY): Condensed from "Producing Your Own Renewable Energy" by Andy Karnitz, published in Humboldt Coalition for Property Rights Newsletter (Summer 09)

13, 14. Renewable Energy Institute: http://www.RenewableEnergyInstitute.org

15. http://ensia.com/voices/its-time-to-rethink-americas-corn-system/

16. http://www.nrdc.org/air/transportation/hydrogen/hydrogen.pdf

17. http://www2.epa.gov/hydraulicfracturing

18. http://www.internationalrivers.org/environmental-impacts-of-dams

19. Scientific American, "China's Three Gorges Dam: An Environmental Catastrophe?" Author: Mara Hvistendahl http://www.scientificamerican.com/article.cfm?id=chinas-three-gorges-dam-disaster

20. http://cdn.globalccsinstitute.com/sites/default/files/publications/85741/global-status-ccs-january-2013-update.pdf

21. http://www.sierraclub.org/coal/coal101/faq.aspx

22. http://www.epa.gov/airquality/powerplanttoxics/powerplants.html

23. http://www.world-nuclear.org/info/Country-Profiles/Countries-G-N/Germany/

24. http://www.world-nuclear.org/info/inf17.html

Sidebar

(c) (Ten Strikes Against Nuclear Power): "Forget Nuclear," by Amory B. Lovins, Imran Sheikh, and Alex Markevich of the Rocky Mountain Institute: http://www.rmi.org/sitepages/pid467.php

25. Christopher Flavin, "Low-Carbon Energy," see note 1

26. Amory B. Lovins, commissioned paper, "Energy End-Use Efficiency," accessed Aug 09 at http://www.rmi.org/images/PDFs/Energy/E05-16_EnergyEndUseEff.pdf

27. Jeff Goodell, "Look West, Obama," Rolling Stone (Feb 19, 09)

Part Four: A JUST SOCIETY

Cynthia McKinney quoted from her acceptance speech (GA Congress) in Nov 2004.

Eleanor Roosevelt quoted from remarks at a UN presentation, 1958.

E.O. Wilson quoted from The Creation: An Appeal to Save Life on Earth (2006), p. 75

1. http://www.fao.org/infographics/pdf/FAO-infographic-SOFI-2012-en.pdf

2. WorldHunger.org, 2013 World Hunger and Poverty Facts and Statistics http://worldhunger.org/articles/Learn/world%20hunger%20facts%202002.htm#Number_of_hungry_people_in_the_world

3. End Poverty 2015 Millennium Campaign, press release (June 23, 09): http://www.endpoverty2015.org

4. Michael Renner, "Environment a Growing Driver in Displacement of People," Worldwatch news reports (Sept 17, 08)

5. Overcoming Barriers: Human Mobility and Development, Human Development

   Report 2009, UN Development Program, http://hdr.undp.org

6. "Population Challenges: The Basics," published by The Population

Institute and available as a downloadable pdf from their website: http://www.populationinstitute.com.

7. Robert Engleman, "Population and Sustainability: Can We Avoid Limiting the Number of People?" Scientific American (June 09); this is also the source of the boxed text in the right margin: http://www.scientificamerican.com/article. cfm?id=population-and-sustainability. Engleman is VP for programs at the Worldwatch Institute and author of More: Population, Nature, and What Women Want (Island Press). See current numbers on the federal government's "population clocks" site: http://www.census.gov/main/www/popclock.html

8. Nicholas D. Kristof and Sheryl WuDunn, "Why Women's Rights Are the Cause of Our Time," The New York Times Magazine (Aug 29, 09)

9. http://endviolence.un.org/index.shtml

10. http://www.worldwatch.org/node/5847

11. Robert Engleman, Scientific American, see note 7 above

12. Optimum Population Trust and the London School of Economics, "Fewer Emitters, Lower Emissions, Less Cost," http://www.optimumpopulation.org/ reducingemission.pdf

13. Hillary Clinton quoted by interviewer Mark Landler in "A New Gender Agenda," The New York Times Magazine (Aug 29, 09)

14. United Nations, "Human Rights and Poverty: 60 Years Later," http://www.un.org/works/sub3.asp?lang=en&id=63

15. http://hdr.undp.org/en/reports/global/hdr2013/

16. Bonney Hartley, UN Permanent Forum on Indigenous Issues. "MDG Reports and Indigenous Peoples: A Desk Review," No. 3 (Feb 08): http://www.mdgmonitor.org/factsheets.cfm

17. "Resource Kit on Indigenous Peoples' Issues," UN Permanent Forum on Indigenous Issues, 2008: http://www.un.org/esa/socdev/unpfii/.../resource_kit_indigenous_2008.pdf

18. UN Declaration on the Rights of Indigenous Peoples, Article 20.1: http://www.un.org

19. Bonney Hartley, see note 16

20. Seventh Generation Fund newsletter, March 06: http://www.7genfund.org

21. http://www.ienearth.org/what-we-do/tar-sands/

22. Nellis Kennedy and Winona LaDuke, "Opportunity knocks but it's not Desert Rock," Navajo Times (June 18, 09): http://www.honorearth.org/news/opportunity-knocks it039snot-desert-rock

23. "White Earth Land Recovery Project Receives…Award," Jessie Smith Noyes Foundation press release: http://www.foundationcenter.org/pnd/news/

24. https://www.adbusters.org/blogs/adbusters-blog/occupywallstreet.html

25. http://www.huffingtonpost.com/2008/03/18/obama-race-speech-read-th_n_92077.html

26. In These Times, "20 Questions with Grace Lee Boggs," http://www.inthesetimes.com/community/20questions/4060/grace_lee_boggs/

27. Congressional Black Caucus Foundation, African Americans and Climate Change: An Unequal Burden (2004): http://www.cbcfinc.org

28. Sunny, Sanwar (2011). Green Buildings, Clean Transport and the Low Carbon Economy: Towards Bangladesh's Vision of a Greener Tomorrow.

29. Hayes Morehouse, Ella Baker Center for Human Rights blog (Aug 21, 09): http://www.ellabakercenter.org/blog

30. http://www.pewhispanic.org/2013/02/15/u-s-immigration-trends/ph_13-01-23_ssimmigration_04_increase/

31. http://www.nclr.org/index.php/about_us/

32. Denver-based Sisters of Color for Education: http://www.sistersofcolor.org

33. http://www.pewhispanic.org/2013/02/15/u-s-immigration-trends/ph_13-01-23_ss_immigration_04_increase/

34. Charu Chandrasekhar, "Flying While Brown: Federal Civil Rights Remedies to Post-9/11 Airline Racial Profiling of South Asians," Asian Law Journal, Vol 10 (May 03)

35. Asian American Justice Center: http://www.advancingequality.org

36. Asian Pacific Environmental Network: http://apen4ej.org/building.htm

37. The Corps Network: http://www.corpsnetwork.org/index.php?option=com_content&view=article&id=87&Itemid=54

Sidebar

(a) (The Seven Foundations of a Just, Sustainable World): Ellis Jones, Ross Haenfler, and Brett Johnson, The Better World Handbook (2007)

Part Five: ECONOMICS

1. Peter M. Senge, Joe Laur, Bryan Smith, Nina Kruschwitz, and Sara Schley, The Necessary Revolution (2008). Excerpt, "Seeing the Whole Picture" in Sustainable Systems at Work, Discussion Course by Northwest Earth Institute, 2009: http://www.nwei.org

2. Peter Senge et al., The Necessary Revolution (2008). List slightly abridged for space.

3. David Korten, "Living Wealth: Better Than Money," Yes! Magazine (Fall 07): http://www.yesmagazine.org/issues/stand-up-to-corporate-power/1834

4. The New Economy Working Group: http://www.neweconomyworkinggroup.org

5, 6. David Korten, "Living Wealth," see note 3 above

7. The New Economy Working Group (in process)

8. http://www.huffingtonpost.com/2013/10/09/richest-1-percent-wealth_n_4072658.html

9. Laura Carlsen, "Americans Policy Report the Mexican Farmers' Movement: Exposing the Myths of Free Trade" (2003): http://www.ifg.org/analysis/wto/cancun/mythtrade.htm

10. Citizens Trade Campaign 2009 "Trade Reform, Accountability, Development and Employment (TRADE) Act" had 75 US Congressional co-sponsors in 2008.

11. Hazel Henderson, "Redefining Economic Growth and Reshaping Globalization Toward Sustainability 2009," International Conference on Concerted Strategies to Meet the Environmental and Economic Challenges of the 21st Century. Vienna, Austria

(April 2009): http://www.clubofrome.org/eng/meetings/vienna_2009/presentations.asp

12. List compiled from the following sources: Henderson (2002, 2003), Jubilee USA, Redefining Progress, Global Policy Forum, and the Global Marshall Plan Initiative's "Five Elements of an Eco-Social Economy"

13. Rachel Dixon, "'Teach us how to fish—do not just give us the fish'; Does buying Fairtrade products really make a difference to people's lives?" http://www.guardian.co.uk/environment/2008/mar/12/ethicalliving.lifeandhealth/print

14. Norwegian Nobel Committee, Oct 06: http://www.grameen-info.org/index.php?option=com_content&task=view&id=197&Itemid=197

15. David Suzuki with Faisal Moola, "Life-altering planetary experience," Science Matters (Oct 2, 09), email newsletter

16. David Korten, "Why This Crisis May Be Our Best Chance to Build a New Economy," YES! Magazine (Summer 09)

17. Chris Maser explains, in detail, in Earth in our Care (2009)

18. Janine Benyus, Biomimicry: Innovation Inspired by Nature (1997)

19. Definition of sustainable development: 1987, United Nations World Commission on Environment and Development

20. Joyce Marcel, "Seventh Generation buys itself," Vermont Business Magazine (July 1, 2000): http://findarticles.com/p/articles/mi_qa3675/is_200007/ai_n8903496/

21. Hazel Henderson, Ethical Markets: Growing the Green Economy (2006)

22. John R. Ehrenfeld, "The Roots of Sustainability," MIT Sloan Management Review (Winter 05). IN Sustainable Systems at Work, Discussion Course 2009, Northwest Earth Institute

23. John Elkington, Chapter 1 Enter the Triple Bottom Line (Aug 17, 04): http://www.johnelkington.com/TBL-elkington-chapter.pdf

24. John Talberth, "A New Bottom Line for Progress." Chapter 2, SOTW 08 by the Worldwatch Institute.

25. Earth Charter International. "The Earth Charter, GRI, and the Global Compact: Guidance to Users on the Synergies in Application and Reporting" (2008), (c) Global Reporting Initiative, downloadable pdf from http://www.globalreporting.org

26. RiskMetrics Group, "Corporate Governance and Climate Change: Consumer and Technology Companies." A Ceres Report (Dec 08): http://www.ceres.org/Page.aspx?pid=1002

27. Carl Frankel, "Putting the Brakes on Fast Money: An Interview with Woody Tasch," Chronogram Magazine (April 27, 09): http://www.chronogram.com/issue/2009/5/Community+Notebook/Putting-the-Brakes-on-F...8/27/2009

28. See also Sourcewatch wiki: http://www.sourcewatch.org

29. http://www.ChevronToxico.com

30. http://www.ftc.gov/news-events/media-resources/truth-advertising/green-guides

31. Larry Lohmann, "When Markets are Poison: Learning about Climate Policy from the Financial Crisis" (Sept 09): http://www.thecornerhouse.org.uk/pdf/briefing/40poisonmarkets.pdf

32. Stacy Mitchell, "The Corporate Co-Opt of Local" (July 9, 09): http://www.newrules.org/retail/article/corporate-coopt-local

33. Stacy Mitchell, "Local Where? Big corporations are finding ways to sell themselves as the folks next door" (July 22, 09): http://www.7dvt.com

Sidebar

(a) Ellen Brown, "The Public Option in Banking: How We Can Beat Wall Street at its Own Game" (Aug 5, 09): http://www.webofdebt.com/articles/public_option.php; Ellen Brown, "But Governor, You Can Create Money! Just Form Your Own Bank" (May 26, 09): http://www.webofdebt.com/articles/but_governor.php

34. Chris Maser, from "True Community is Founded on a Sense of Place, History and Trust" (2008), Social Essay 28; and from "The Commons Usufruct Law" (2009), Social Essay 31: http://www.chrismaser.com

35. Michael Shuman, The Small-Mart Revolution: How Local Businesses Are Beating the Global Competition (2006)

36. Michael Shuman, "Local Stock Exchanges and National Stimulus," posted on Sept 2, 09: http://www.small-mart.com/home

37. "Resilience of the co-operative business model in times of crisis" and statistical information on the International Co-operative Alliance website: http://www.ica.coop/coop/statistics.html

38, 39. "Our Story" about Development Deposits: http://www.shorebankcorp.com

Sidebar

(b) (Choose Well): Chris Maser, "True Community is Founded on a Sense of Place, History and Trust" (2008), Social Essay 28: http://www.chrismaser.com

Part Six: COMMUNITY

1. Phone interview with Bruce Davidson in Shutesbury, MA, Sept 23, 09

2. Bill McKibben, "A Place that Makes Sense: On Not Living Too Large," Sept 23, 08: http://www.christiancentury.org/article.lasso?id=5225

3. Stephanie Hemphill, "Swedish Town Takes Sustainability to New Level," Minnesota Public Radio, Morning Edition (July 8, 09): http://minnesota.publicradio.org/display/web/2009/07/07/sustainable_sweden/

4. Alison Pruitt, "Hammarby Sjöstad, Stockholm Becomes Model of Sustainability," July 13, 09: http://www.energyboom.com/policy/Hammarby_Sjostad_Stockholm_Becomes_Model_of_Sustainability

5. http://www.epa.gov

6. "The State of Green Building," a White Paper, published by ThermaTru: http://www.thermatru.com/pdfs/WhitePaper.pdf

7. http://www.environmentamerica.org/uploads/qk/zy/qkzycNV75kmR8g8HIAR7rw/AME_BBA_web.pdf

8. Global Footprint Network expands on the carbon footprint concept: http://www.footprintnetwork.org

9. Pat Murphy, "The Energy Impact of Our Buildings," http://www.communitysolution.org

10. Andrew Michler, "Seven steps to a sustainable building, a performance path" (April 16, 09): http://www.igreenbuild.com

11. David Owen, "Is Manhattan one of the greenest cities around?" The New Yorker: http://www.newyorker.com/online/blogs/newsdesk/2009/09/david-owen-green-metropolis.html

12. http://www.whiteroofproject.org

13. "Green walls—go vertical!": http://www.eltlivingwalls.com/; http://www.g-sky.com/; http://www.cnn.com/2009/TECH/science/06/28/green.walls/index.html; http://www.mnn.com/business/commercial-building/blogs/pnc-unveils-six-story-green-wall; http://www.mnn.com/the-home/building-renovating/stories/green-walls-of-china; http://www.treehugger.com/files;/2008/09/11-buildings-wrapped-in-green-walls.php

14. http://www.igreenbuild.com/cd_3218.aspx

15. http://www.epa.gov/climatechange/ghgemissions/gases/co2.html

16. Toronto Transportation Services, Idling Control Bylaw: http://www.toronto.ca/transportation/onstreet/idling.htm

17. http://www.worldbank.org/en/news/feature/2013/01/31/what-next-sustainable-transport-cities

18. "Redesigning Urban Transport," Earth Policy News, adapted from Ch. 10, "Designing Cities for People," in Lester R. Brown, Plan B 3.0: Mobilizing to Save Civilization (2008)

19. http://www.railway-technology.com/features/feature124824/

20. http://www.railjournal.com/index.php/high-speed/chinas-high-speed-programme-back-on-track.html

21. See note 18

22. Lloyd Alter, "They are Building Bicycle Superhighways in Copenhagen" (Aug 21, 09): http://www.treehugger.com/files/2009/08/copenhagen-bicycle-superhighways.php

23. See note 18

24. Rocky Mountain Institute, "Transformational Trucking, How the Trucking Industry Can Avoid the Automotive Industry's Fate": http://www. treehugger.com/files/2009/03/transformational-how-trucking-industry-avoid-automotive-fate.php

25. Sierra Club: http://www.SierraClub.org

26. http://www.uctc.net/access/38/access38_carsharing_ownership.pdf

Part Seven:
GETTING PERSONAL

1. The Hunger Project: http://www.thehungerproject.org; Food and Agriculture Organization of the United Nations: http://www.fao.org

2. http://www.localharvest.org/buylocal.jsp

3. Ecology Letters *Intensive Agriculture Erodes Beta-Diversity at Large Scales.* Daniel S. Karp, Andrew J. Rominger, et al. 2012

4. http://www.sierraclub.org/factoryfarms/factsheets/antibiotics.asp

Sidebar

(a) (The Six Principles of Food Sovereignty): http://www.foodsovereignty.org

5. http://www.epa.gov/owow/NPS/Ag_Runoff_Fact_Sheet.pdf

6. Billions in Farm Subsidies Underwrite Junk Food: http://www.huffingtonpost.com/2011/09/22/farm-subsidies-junk-food_n_975711.html

7. http://www.mnn.com/food/healthy-eating/stories/monsanto-wins-supreme-court-case-on-gmo-soybean-seeds

8. http://thinkprogress.org/health/2013/10/17/2787921/african-countries-join-anti-monsanto-protests/

9. http://www.worldwatch.org/agriculture-and-livestock-remain-major-sources-greenhouse-gas-emissions-0

10. Food and Agriculture Organization of the United Nations, "Livestock's Long Shadow: Environmental Issues and Options," Rome (2006), p. 272

11. Gidon Eshel and Pamela Martin, Univ of Chicago study (2006), "Vegan Diets Healthier for Planet, People Than Meat Diets," Environmental Science and Technology (2008)

12. Journal of the American Dietetic Association (June 03), Vol 103, No 6: pp. 748–65

13. http://www.animalwelfareapproved.org

14. http://journeytoforever.org/farm_library/balfour_sustag.html

15. http://www.nwf.org/How-to-Help/Garden-for-Wildlife/Certify-Your-Wildlife-Garden.aspx?campaignid=WH12L1CSWWX&adid=45625

16. http://www.epa.gov/HPV/pubs/general/hazchem.htm

17. http://www.yesmagazine.org/issues/rx-for-the-earth/828

18. http://www.cancer.org/cancer/cancerbasics/lifetime-probability-of-developing-or-dying-from-cancer

19. http://blog.saferchemicals.org/2011/02/top-tips-to-keep-toxic-chemicals-at-bay.html

20. http://echa.europa.eu/regulations/reach/understanding-reach

21. http://europa.eu/legislation_summaries/consumers/product_labelling_and_packaging/co0013_en.htm

22. http://www.ncbi.nlm.nih.gov/pubmed/11999798

23. http://www.toxicfreefiresafety.org

24. http://www.footprintnetwork.org/en/index.php/GFN/page/world_footprint/

25. http://worldcentric.org/conscious-living/expanding-eco-footprint

26. Juliet Schor, "Forget commercialism! The new realities of consumption and the economy" (posted Nov 18, 08); see also "The Politics of Consumption," Boston Review (Summer 99): http://www.simpleliving.net

Sidebar

(b) Fowler, James W. (1981). Stages of Faith, Harper & Row

Leah Beck